AIR, WATER AND SOIL POLLUTION SCIENCE AND TECHNOLOGY

GREEN PLANTS AND POLLUTION: NATURE'S TECHNOLOGY FOR ABATING AND COMBATING ENVIRONMENTAL POLLUTION

AIR, WATER AND SOIL POLLUTION SCIENCE AND TECHNOLOGY

Trends in Air Pollution Research
James, V. Livingston (Editor)
2005. ISBN: 1-59454-326-7

Agriculture and Soil Pollution: New Research
James, V. Livingston (Editor)
2005. ISBN: 1-59454-310-0

Air Pollution: New Research
James, V. Livingston (Editor)
2007. ISBN: 1-59454-569-3

Air Pollution Research Advances
Corin G. Bodine (Editor)
2007. ISBN: 1-60021-806-7

Water Pollution: New Research
A.R. Burk (Editor)
2008. ISBN: 1-59454-393-3

Marine Pollution: New Research
Tobias N. Hofer (Editor)
2008. ISBN: 978-1-60456-242-2

Complementary Approaches for Using Ecotoxicity Data in Soil Pollution Evaluation
M. D. Fernandez and J. V. Tarazona
2008. ISBN: 978-1-60692-105-0

Complementary Approaches for Using Ecotoxicity Data in Soil Pollution Evaluation
M. D. Fernandez and J. V. Tarazona
2008. ISBN: 978-1-60876-411-2 (E-book)

Lake Pollution Research Progress
Franko R. Miranda and Luc M. Bernard (Editors)
2008. ISBN: 978-1-60692-106-7

Lake Pollution Research Progress
Franko R. Miranda and Luc M. Bernard (Editors)
2008. ISBN: 978-1-60741-905-1 (E-book)

Heavy Metal Pollution
Samuel E. Brown and William C. Welton (Editors)
2008. ISBN: 978-1-60456-899-8

Heavy Metal Pollution
Samuel E. Brown and William C. Welton (Editors)
2008. ISBN: 978-1-61668-049-7 (E-book)

Air Pollution and Ship Emissions
Jacob Boutin *(Editor)*
2010. ISBN: 978-1-60876-087-9

**Heavy Metal Compounds in Soil:
Transformation upon Soil
Pollution
and Ecological Significance**
*Tatiana M. Minkina, Galina V.
Motusova, Olga G. Nazarenko
and Saglara S. Mandzhieva*
2010. ISBN: 978-1-60876-466-2

**From Soil Contamination to
Land Restoration**
Claudio Bini
2010. ISBN: 978-1-60876-853-0

Green Plants and Pollution
Rajiv Sinha
2010. ISBN: 978-1-61668-147-0

Green Plants and Pollution
Rajiv Sinha
2010. ISBN: 978-1-61668-403-7
(E-book)

AIR, WATER AND SOIL POLLUTION SCIENCE AND TECHNOLOGY

GREEN PLANTS AND POLLUTION NATURE'S TECHNOLOGY FOR ABATING AND COMBATING ENVIRONMENTAL POLLUTION

RAJIV K. SINHA

AND

SHWETA SINGH

Nova Science Publishers, Inc.
New York

12-78441

DEC 2 2 2010

For permission to use material from this book please contact us:
Telephone 631-231-7269; Fax 631-231-8175
Web Site: http://www.novapublishers.com

NOTICE TO THE READER

The Publisher has taken reasonable care in the preparation of this book, but makes no expressed or implied warranty of any kind and assumes no responsibility for any errors or omissions. No liability is assumed for incidental or consequential damages in connection with or arising out of information contained in this book. The Publisher shall not be liable for any special, consequential, or exemplary damages resulting, in whole or in part, from the readers' use of, or reliance upon, this material.

Independent verification should be sought for any data, advice or recommendations contained in this book. In addition, no responsibility is assumed by the publisher for any injury and/or damage to persons or property arising from any methods, products, instructions, ideas or otherwise contained in this publication.

This publication is designed to provide accurate and authoritative information with regard to the subject matter covered herein. It is sold with the clear understanding that the Publisher is not engaged in rendering legal or any other professional services. If legal or any other expert assistance is required, the services of a competent person should be sought. FROM A DECLARATION OF PARTICIPANTS JOINTLY ADOPTED BY A COMMITTEE OF THE AMERICAN BAR ASSOCIATION AND A COMMITTEE OF PUBLISHERS.

LIBRARY OF CONGRESS CATALOGING-IN-PUBLICATION DATA

Green plants and pollution / author, Rajiv Sinha.
    p. cm.
Includes bibliographical references and index.
ISBN 978-1-61668-147-0 (softcover)
1. Pollution. 2. Herbaceous plants--Environmental aspects. 3. Water--Purification--Biological treatment. 4. Air--Purification. 5. Air quality management. I. Title.
TD177.S565 2010
363.73'6--dc22

2010001749

Published by Nova Science Publishers, Inc. † New York

CONTENTS

PREFACE

Green plants work as a 'natural pollutant sink' on earth intercepting dust and pollutants from air and water. During photosynthesis, plants absorb CO_2 and simultaneously other gases like oxides of sulfur & nitrogen (SO_2 & NO_x), ozone (O) and airborne ammonia (NH_3) through their stomata. Plants also reduce air pollution by intercepting suspended particulate matters (SPM) and aerosols and retaining them on the leaf surface. Trees take up more pollutants including the particulate pollutants ($PM10$), than shorter vegetation. This book presents new and significant research in the role of plants in protecting the environment.

ACKNOWLEDGMENT

We are grateful to all those learned authors, editors and publishers of books which provided valuable information on the subject and helped in the preparation of this volume. Their names have been duly referred in the list of references. They have great expertise in plants combating air and water pollutants which is an emerging area of studies all over the world as the problem of environmental pollution grows unabated. We especially acknowledge the learned editors Dr. S.N. Singh & Dr. R.D. Tripathi and all the learned authors of the book '*Environmental Bioremediation Technologies*'; Springer Publication, New York (2006) which provided valuable information on the subject for completion of the book.

And, to the best of our knowledge we have taken all care not to violate the copyrights of the learned authors but if that might have had happened unknowingly & unintentionally, we all duly apologize to those learned authors and scientists. Our intention is to spread the knowledge about 'the role of mother nature' (all green plants - trees and forests) in combating air and water pollution' among the global human society so as to worship mother nature and have reverence for her, and also to make aware the policy makers and developers to protect forests and green areas in cities which protects the human society.

We express our deep gratitude to Sir Frank Columbus Editor-in-Chief, NOVA Science Publishers, USA and his entire NOVA team for publishing this chapter in this book. This is a great recognition for authors.

Rajiv K. Sinha
Shweta Singh

ABSTRACT

Green plants work as a 'natural pollutant sink' on earth intercepting dust and pollutants from air and water. The National Botanical Research Institute, India, has identified over 50 plants as 'natural pollutant sink'. During photosynthesis, plants absorb CO_2 and simultaneously other gases like oxides of sulfur & nitrogen (SO_2 & NO_x), ozone (O) and airborne ammonia (NH_3) through their stomata. Plants also reduce air pollution by intercepting suspended particulate matters (SPM) and aerosols and retaining them on the leaf surface. Trees take up more pollutants including the particulate pollutants (PM_{10}), than shorter vegetation. Study indicated 27 % reduction in dust particles in Hyde Park, London, by green cover of 2.5 km^2. Over 300 indoor plants have been identified that has the wonderful property of absorbing and removing indoor air pollutants such as benzene, formaldehyde, trichloroethylene and all other VOCs.

Some plants are relatively more 'tolerant' to air pollutants and can bio-accumulate them in their cells and tissues. Others are 'sensitive' which gets injured and respond to their injury by some visible morphological & physiological changes. Such plants work as 'bio-indicator' of air pollutants and as an 'early warning device' for providing information on deteriorating air quality in a region.

Several aquatic plants have been found to absorb and detoxify chemical pollutants including heavy metals and organic pollutants from the water bodies. Plant enzymes and the symbiotic microbes on their roots plays important role in biodegradation of complex organics. The floating hydrophyte water hyacinth (*Eichhornia cressipes*) can remove heavy metals by 20-100 %. In just 24 hours the weed can extract more than 75 % of lead (Pb) from polluted water. It also absorbs cadmium (Ca), nickel (Ni), chromium (Cr), zinc

(Zn), copper (Cu), iron (Fe) and pesticides and several toxic chemicals. Another freshwater species *Ceratophyllum demersum* can bio-accumulate arsenic (As) from water with a 20,000 – fold concentration factor.

Some plants (both terrestrial and aquatic) have also been found to remove radio-nuclides e.g. uranium (U), strontium (Sr^{90}) and cesium (Cs^{137}) from polluted water. Study showed that the sunflower plants grown hydroponically in the pond could take up 90 % of the cesium-137 (Cs^{137}) (from 80 Bq/L of Cs^{137}) in just 12 days. It also reduced strontium – 90 (Sr^{90}) concentrations from 200 μg/L to 35 μg/L within 48 hours which was further reduced to 1 μg/L. Genetic engineering has produced some 'wonder transgenic plants' to combat air & water pollution.

Keywords: Plants as Living Biofilter & Natural Pollutant Sinks; Plants as Bioindicators of Pollutants in Air & An Early Warning Devices; Transgenic (Genetically Engineered) Plants Combating Air Pollution; Plants Combating Indoor Air Pollutants;; Hydroponically Grown Terrestrial Plants Combating Water Pollutants; Rhizofiltration of Pollutants by Plant Roots; Phytovolatalization of Pollutants by Leaves; Constructed Wetlands Technology for Treatment of Wastewater & Polluted Water.

INTRODUCTION

All green plants (the terrestrial and aquatic vegetation on earth) work as a 'natural pollutant sink' on earth intercepting dust and pollutants from air and water. This fact has been long recognised especially for the greenhouse gas carbon dioxide (CO_2) which is sequestered by green plants in the process of photosynthesis. Once the pollutants are released into the atmosphere, only the plants are the hope, which can mop up the pollutants by adsorbing them on their leaf surfaces or absorbing and metabolising (degrading) them into their metabolic system. Plants play an important role in mitigation of highly polluted atmosphere and extreme climates in urban and semi-urban areas. Yang et al. (2005) (mentioned in Singh & Verma (2006) in a study on the role of urban forest and vegetation in air pollution abatement & reduction, found that there was 1262.4 tons of pollutants reduction by the forest cover in Beijing. Planting of trees and shrubs was recommended as a way to combat dust pollution in Russian cities and there was 2-3 times reduction in dust fall by planting a 8 m wide green belt between the roads and buildings.

During photosynthesis, plants absorb CO_2 and simultaneously other gases like oxides of sulfur & nitrogen (SO_2 & NO_x), ozone (O) and airborne ammonia (NH_3) through their stomata. (Bergmann, 1995; Yunus & Iqbal, 1996). Simonich and Hites (1994) reported that plants are also an efficient sink for polyaromated hydrocarbons (PAHs) in the air. Once inside the leaf, gases diffuse into the spaces between the cells of the leaf to be absorbed by water films or chemically altered by the plant enzymes in the metabolic process. Plants also reduce air pollution by intercepting suspended particulate matters (SPM) and aerosols and retaining them on the leaf surface by process of 'dry deposition'. Pollutant removal rates are the highest when the vegetative surfaces are wet or damp (such as after the rains or morning dews in winter).

Such conditions can increase pollution removal rates ten-fold because the entire tree surface (barks, leaves & branches) is available as pollutant sink.

The 21st century has been called as the 'Century of Water' and the year 2000 was declared as the 'International Year of Freshwater' by UNEP. Water - the 'elixir of life', is the thread that knits together the web of life on earth. It purifies and keeps our bodies healthy, provides us with food, is home to millions of creatures, regulates the global climate, dilute pollutants, and sustains every nation's economic wealth as the essential resource for our industries, agriculture and transportation.

Chemical pollution of water is a growing concern all over the world. Chemicals are released into the global aquatic systems in the form of liquid, dust, fumes or gas. Such releases can be planned (part of the development process e.g. industrial smokestack emissions, automobile exhaust emissions which settles down in rivers and lakes, discharge of domestic and industrial wastewater into rivers and streams, etc.) or unplanned (accidental). Chemicals can also enter the water system during transport (e.g. from the site of manufacture to the site of use), during their intended use (e.g. pesticide spray) or through disposal in landfills and waterways.

The notable environmental contaminants of water bodies are 'radionuclides' as well as 'organic' and 'inorganic' pollutants. Several inorganic contaminants in fact also constitute the 'micro & macro nutrients' for aquatic organisms in traces. Inorganic pollutants include nitrate (N), phosphate (P), per chlorate, cyanide (CN), boron (B), copper (Cu), iron (Fe), manganese (Mn), molybdenum (Mo), zinc (Zn), arsenic (As), cobalt (Co), chromium (Cr), nickel (Ni), selenium (Se), vandalium (V), fluoride (F) and strontium (Sn) etc.

Inorganic elements such as boron, copper, iron, manganese, molybdenum and zinc are essential as plant nutrients in traces but become pollutants when present in excess. Inorganic elements such as arsenic, cobalt, iron, manganese, zinc, chromium, nickel, selenium, vandalium, fluoride and strontium are essential as nutrients to aquatic animals in traces but become pollutants when present in excess. There are some most toxic trace elements which are NOT required by any organisms such as lead (Pb), cadmium (Cd) and mercury (Hg).

Chapter 1

PLANTS COMBATING AIR POLLUTION

Plants play an important role in combating air pollution and working as a 'natural pollutant sink'. The broad leaved shrubs and trees intercept considerable amount of dust and pollutants from the air. Leaf surfaces are most efficient at removing pollutants that are water-soluble including SO_2, NO_2 and O_3.

CHEMICALS IN THE AIR

Millions of tons of carbon dioxide, sulfur dioxide and nitrogen oxides are released into the atmosphere every day by our industries and automobiles. Ambient air today contain a frightening mix of toxic chemicals like carbon monoxide (CO), oxides of sulfur (SOx) and nitrogen (NOx), reactive hydrocarbons (HC) also known as volatile organic carbons (VOCs), total suspended particulate matter (TSP), heavy metal lead and organic compounds resulting from the automobiles. Five major air pollutants account for 98 % of pollution: carbon monoxide (52 %), sulfur oxides (14 %), VOCs (14 %), particulate matters (4 %) and nitrogen oxides (14 %). The remainder consists of lead, which is down 90 % since 1983. Most gaseous air pollutants are totally transparent except nitrogen oxide (NO_2), which is brown.

The oxides of nitrogen (NO_x) and the volatile organic carbons (VOCs) undergo photochemical reaction in sunlight to produce more deadly secondary pollutants called 'tropospheric ozone'(O_3) and the 'periacetyl nitrate' (CH_3COONO_2). Ozone and PAN are two deadly components of this 'photochemical urban smog' which plagues several cities of world today. Smog contains chemical cocktail of deadly pollutants from the automobiles-about 10,000 to 30,000 ppm of carbon monoxide (CO), 100 to 400 ppm of nitric oxide (NO_2), 600 to 3000 ppm of hydrocarbons (HCs) or the volatile organic compounds (VOCs), 50 to 150 ppm of ozone (O_3) and 50 to 250 ppm of periacetyl nitrate (PAN), sulfur dioxide (SO_2), and the suspended particulate matters (SPM). Aldehydes are major products in these reactions. Formaldehyde and acrolein account for about 50 % and 5 %, respectively, of the total aldehyde in urban air. Acrolein is a hazardous air pollutant (HAP).

PLANTS AS NATURAL POLLUTANT SINK

The National Botanical Research Institute (NBRI), Lucknow, India, has identified over 50 plants (trees, herbs and shrubs) as 'natural pollutant sink' (Anonymous, 1986). Some of the very prominent ones are –

Allium cepa
Agremone mexicana
Azadirachta India
Avena sativa
Bougainvallea spp.
Coleus blumei
Cucumis sativus
Ficus bengalhensis
Ficus religiosa
Gingko biloba
Mangifera indica
Nerium odorum
Phaseolus vulgaris
Phoenix sylvestris
Sachharaum officinarum
Saraca indica
Simmondsia chinesis
Triticum aestivum
Zea mays

There are unconfirmed reports about some trees e.g. *Ginkgo biloba* (very ancient and a living fossil), *Ficus religiosa* and the *Ficus bengalhensis* as having metabolic properties to assimilate carbon monoxide (CO) along with CO_2 in photosynthesis. (Verbal communication by Late Prof. GS Nathawat of

Department of Botany & Environmental Sciences, University of Rajasthan, Jaipur, India, as having appeared in *Canadian Journal of Botany* in the 1980s).

(No wonder then, these trees are mentioned in Hindu & Buddhist religious epics as 'divine trees' with tremendous ability to purify the air in the surroundings. *Gingko* has been traditionally planted in Buddhist monastries and the other two trees *F. religiosa* & *F. bengalhensis* are trees of preference in all hindu temples. Hindu scriptures even mention the *Ficus religiosa* tree as the abode of Lord Krishna and hence worshipped).

IMPACT OF POLLUTANTS ON PLANTS AND PLANTS AS BIO-INDICATOR OF TOXIC POLLUTANTS IN AIR : AN EARLY WARNING DEVICE OF DETERIORATING AIR QUALITY

Air pollutants like nitrogen oxide (NO_2), sulfur oxide (SO_2), tropospheric ozone (O_3) and the suspended particulate matters (SPMs) have pernicious effects of varying magnitudes on important crops like wheat, mustard, mung and spinach plants at higher concentrations and on prolonged exposures. Some plants are, however, relatively more 'tolerant' to air pollutants and can bio-accumulate them in their cells and tissues. Others are 'sensitive' which gets injured and damaged and respond to their injury by way of some visible morphological as well as physiological and anatomical changes. The more sensitive ones may altogether 'disappear' in response to pollutants in the air. Such plants work as 'bio-indicator' of air pollutants and work as an 'early warning device' for providing information on deteriorating air quality in a region'

A number of air pollutants can onset early visible damage on plants and they can often provide very first evidence of air pollution. The visible plant damages that work as bioindicator of toxic pollutants in the air include 'mottled foliage', 'burning at leaf tips or margins', 'twig dieback', 'stunted growth' 'premature leaf and flower drop', 'abortion or early drop of blossoms', 'delayed maturity' and 'reduced yield or quality of fruits & seeds'. This provides an inexpensive way to bio-monitor the presence of air pollutants and quality of air in a region.

Impact of Sulphur Dioxide (SO_2)
on Plants & the Bioindicator Species

Common sources of SO_2 are coal power plants, oil refineries, copper & iron smelters & fossil fuel furnaces. SO_2 is unique, at low concentration it is in fact beneficial to plants, while injurious at high concentrations. The injury is both visible and invisible. Many plants are known to be injured by SO_2 under natural & experimental exposure conditions. It has potential to reduce both yield and nutritional quality of crops and more perhaps to the mango crops.

Gaseous SO_2 is highly soluble in water and is ionised to form the hydrogen (H^+), sulfite (SO_3^{2-}) and bisulfite (HSO_3^-) ions depending upon the pH of the plant tissues. Free radicals produced during SO_3^{2-} oxidation, have been known to destroy many physiologically important compounds like amino acids, plant hormone IAA, chorophyll and β carotene in plants. Laboratory experiments have shown significant reductions in non-structural carbohydrates and proteins, and the nitrogen contents of seeds, fruits and vegetables when exposed to SO_2. (Singh & Verma, 2006).

The exposure of succulent, broad-leaved plants to SO_2 usually result in dry, papery blotches, colored tan, straw or even white and turn to interveinal browning on necrosis. More sensitive plants are alfalfa (*Medicago sativa*), beans, beets, buckwheat (*Fagopyrum esculentum*), soybean (*Glycine max*) and sunflower (*Helianthus anus*). At the National Monitoring Network in The Netherlands alfalfa and buckwheat are used for bio-monitoring SO_2 in ambient air. (Posthumus, 1983, 1985).

Lichens as Bio-indicators of Sulfur in the Environment:
A Case Study from UK

Lichens are symbiotic organism resulting from the symbiosis of algae and fungi. They can grow on old walls and on the tree trunks of old trees in parks. Many species of lichens are extremely sensitive to sulfur dioxide (SO_2), and die while concentrations are still quite low. Rich flora of lichens are found in UK in pollution zones of 35 units and under, and disappears above that. Only 'crustose lichens' if any, can survive in heavy pollution zone of 150 units. Crustose lichens and few 'foliose lichens' survive in 70 – 125 units of pollution. 'Beared lichens' are occasionally found in low pollution zone of 40 – 50 units.

In U.K. such lichens vanished from most industrial areas during the 19th century. As the air became cleaner after the legislation for control of sulfur dioxide pollution came into effect, the lichens have now returned to their habitat.

Impact of Nitrogen Oxides (NO$_2$) on Plants

NO$_2$ is less disruptive to plants and rather work as 'nutrient' in small concentrations but are phytotoxic at higher levels and prolonged exposures. Low NO$_2$ has been found to induce the production of chlorophyll pigments while it is reduced at higher concentrations. In sunflower, 300 ppb of NO$_2$ exerted a nutritional effect on plant growing on nitrogen-deficient soils, while 2000 ppb was phytotoxic. Study also revealed that although NO$_2$ was stimulant to growth at low concentrations, it was damaging to plants in combination with SO$_2$ at the same concentration.

Impact of Suspended Particulate Matters (SPM) on Plants

Suspended particulate matters (SPM) may cause ultrastructural and physiological disturbances in plants. Wax crystals which are the barriers between the plants and the environment fuse and flatten with age, but in the presence of SPM, erosion rate of wax increases. At worst, the epistomatal chamber of the leaf surface may be plugged totally by the withered and fused wax inhibiting transpiration, photosynthesis & respiration with grave consequences for the plants. Dust deposition on leaf cuticle due to particulate penetration into the epicuticular wax may reduce light incidence and reduce net photosynthesis. Dust deposition also can lead to clogging of stomata and inhibiting all physiological functions.

Impact of Tropospheric Ozone (O$_3$) on Plants & the Bioindicator Species

Ozone is a secondary pollutants resulting from photochemical reaction between hydrocarbons (HC) and NO$_x$ from the automobiles in sunlight. It is probably most phyto-toxic air pollutant to plants disrupting photosynthesis and other metabolic functions.

Long-term exposure to near ambient ozone may lead to appearance of 'chlorotic symptoms', 'stippling' and 'speckling' (characterized by numerous tiny dots) on the upper leaf surface. Reduction in crop yield due to disrupted photosynthesis are other symptoms without any visible injury.

Some excellent bioindicator plant species that has been used widely to detect ozone (O_3) in the lower atmosphere are tobacco (*Nicotiana tabacum*) cultivars Bel-W3 which are super sensitive to ozone. The classical ozone symptoms in tobacco cultivar Bel-W3 appear as sharply defined dot-like lesions on the adaxial side of the leaves resulting from the death of group of palisade cells. The cultivar Bel-B is highly ozone tolerant and have been used to biomonitor and control ozone in lower atmosphere. Susceptible tobacco plants are injured when the ozone concentration exceed 0.04 ppm. (Upadhaya & Kobayashi, 2006). Morning glory (*Ipomea* spp.) in Japan and the clover plant in Sweden have also been reported as bioindicator plants for ozone. Reduction in growth of radish (*Raphanus sativus*) has been observed as an indicator of ozone in Japan and Egypt. (Izuta et al. 1993).

Impact of Periacetyl Nitrate (PAN) on Plants & the Bioindicator Species

After ozone, PAN is the most phyto-toxic air pollutant. It is also a secondary pollutant in the lower atmosphere resulting by the same photochemical reaction between HC & NO_x which results into ozone formation in sunlight.

PAN is a deadly pollutant for human beings (affecting nervous system) and in plants cause premature senescence and leaf fall. Other symptoms are appearance of bands, blotches, bronzed and silvery areas. Even a minimum exposure for one hour to 0.01 to 0.05 ppm of PAN induces symptoms in susceptible plants.

The bioindicator plants for PAN in the environment are lettuce (*Lactuca sativa*), blue grass (*Poa annua*) and the Swiss chard (*Beta chilensis*). Petunias are also very sensitive to PAN and may be used as indicator plants for the chemical in the ambient air.

Impact of Ethylene on Plants & the Bioindicator Species

Ethylene is also many of the unburnt hydrocarbon products given out from the automobile exhaust. It also results from incomplete combustion of gas, oil and coal. It is a by-product of polyethylene manufacture.

Ethylene can modify the activities of plant growth hormones thus affecting the normal development of plant organs and tissues, without causing leaf tissues damage and necrosis. Injury to broad-leaved plants occurs as the downward curling of leaves and shoots (epinasty), followed by stunting of growth.

Posthumus (1983) studied the use of petunia plant (*Petunia axilliaris*) as the bioindicator plant for ethylene in The Netherlands. *Petunia hybrida* was studied as bioindicator plant for ethylene in Sweden in 1989. Potted petunia plants were placed at distances of 10, 20, 40, 80 and 120 m from a motorway with approximately 30,000 vehicles plying everyday. The results showed that petunia flowers were significantly smaller on plants closer to the motorway than those at distance. The abortion rate of flowers closer to the motorway was more frequent. (Upadhaya & Kobayashi, 2006).

Impact of Fluorides on Plants & the Bioindicator Species

Fluorides are generated from the glass, aluminum, pottery, brick and ceramic industries; from phosphate fertilizer plants, refineries and metal ore smelters. Typical symptoms of fluoride gas or particulate injury are 'yellowish mottle', wavy 'reddish or tan scorching' at the margins or tips of the broad leaved plants or 'tip burn' in grasses and conifers.

Gladiolus (*Gladiolus hortulanus*) is the most widely used plant as a bioindicator and for biomonitoring fluorides.

MECHANISM OF COMBATING AIR POLLUTANTS BY TOLERANT PLANT SPECIES

Several tolerant plant species (trees, shrubs & herbs) have been identified through botanical surveys at the National Botanical Research Institute (NBRI) in Lucknow. Tolerant plant species can bio-accumulate pollutants in their cells and tissues. The response of various antioxidants to automobile exhaust pollutants was studied and it was found that *Amaranthus spinosus* & *Cephalandra indica* were equipped with a very good scavenging system to combat air pollution (Singh & Verma, 2006).

Farooq et al. (1988) exposed 12 common Indian tree species to varying concentrations of SO_2 to determine their tolerance levels and an order of sensitivity was determined like this –

1.Tamarindis indica > 2. Pithecolobiam dulce > 3. Mangifera indica > 4. Ficus rumphie > 5. Holoptelea integrifolia > 6. Bombax ceiba > 7. Ficus bengalansis > 8. Azadirachta indica > 9. Ficus religiosa > 10. Syzygium cumuni > 11. Psidium guajava > 12. Ficus racemosa.

Freer et al. (2004) studied relative pollutant dry deposition velocities and pollutant capture efficiencies of some species widely used in Europe urban and sub-urban areas. This included oak (*Quercus* spp.), alder (*Alnus* spp.), ash (*Fraxinus* spp.), sycamore (*Acer* spp.), Douglas fir (*Pseudotsuga* spp.), weeping fig (*Ficus* spp.) and Eucalyptus (*Eucalyptus globulus*). Species with more complex stem structure and smaller leaves had greater deposition velocities.

Mechanism of Stomatal Closure Protect From Pollutants

Most of the plants when exposed to higher concentrations of air pollutants, tend to restrict the entry of air pollutants by closing their stomata. This is a natural reaction and adaptation in plants to save themselves against toxic situations in the air. Mansfield & Free-Smith (1984) found quick stomatal closure in silver birch in the presence of SO_2 in air. Tomato and peanut also close their stomata quickly on exposure to SO_2 thus restricting their entry.

Ascorbic Acid Provides Defense Mechanism in Plants Against Pollution

High amount of ascorbic acid (Vitamin-C) in plant cells has been implicated in providing resistance to plants against pollutants in air. It also plays important role in protection of chlorophyll from hydrogen peroxide (H_2O_2) induced damage. Ascorbic acid activates the defense mechanism in plants. Due to its multiple role in metabolism and defense of plants, ascorbic acid is used as very reliable parameter to denote tolerance levels in plants against all kinds of stress including pollution stress. Pollutants often increase their phytotoxicity by impinging a decrease in the ascorbic acid contents of plants, which results in increased susceptibility of plants to pollution. Ascorbic acid also has the potential to mitigate the SO_2 induced injury in plants. (Singh & Verma, 2006).

Phyto-assimilation of Sulphur Dioxide (SO₂) by Plants

SO_2 from the atmosphere, finds its entry mainly through the stomata into the leaf cells. Plants can utilize the absorbed SO_2 in a 'reductive sulfur cycle' to form sulfur containing amino acids needed for their growth & development.

SO_2 is converted into SO_3^{2-} and SO_4^{2-} inside the leaf tissues which reacts with organic acids to form amino acids – cysteine, methionine and glutathione. The intermediate compounds formed in the sulphate reduction pathway are APS (Adenyl-5-phosphosulphate), CS^- (Career protein), $CSSO_3$ (Career protein with bound sulphite) and CSS^- (Career protein with bound sulphide). In a study,increased SO_4^{2-} concentrations were found in needles of spruce (*Picea abies*) and Scotspine in Norway when exposed to SO_2 gas. (Singh & Verma, 2006).

Of all plants, the mango trees (*Magngifera indica*) has shown higher tolerance and accumulation capacity to SO_2 in air .It can bioaccumulate high amount of sulfur rich compounds.(Singh & Verma, 2006)

Phyto-assimilation of Nitrogen Oxides (NO₂) by Plants

Direct evidence for the foliar absorption of NO_2 has been obtained by using N^{15} isotopes of nitrogen. Maximum absorption of NO_2 was by three cultivars of hybrid polpar (*Populus* spp.) i.e. 0.3 mg $N/dm^2/d$. Sunflower and maize showed concentrations range between 200 – 1000 ppb.(Okano et al. 1986). NO_2 after entering into the leaves through stomata, dissolves into the intracellular fluid to form nitrous acid which further dissociates into toxic 'nitrites' and H^+ ions. Normally the nitrites gets reduced to ammonia (NH_3) by the enzyme 'nitrite reductase' (NR) and consequently assimilated into amino acid 'glutamate' and proteins, thus alleviating the toxicity while also benefiting the plants. Experiments with N^{15} shows that about 65 % of the absorbed NO_2 is incorporated into organic nitrogen during a 3 hour exposure period in beans.

Phyto-filtration of SPM by Plants

Tree take up more pollutants including the particulate pollutants (PM_{10}), than shorter vegetation. The exposed surface of trees, such as the bark and leaves forms a 'natural sink' for the particulate pollutants, as they provide sites for the gravity or wind-blown settlement of particulates. A study indicated 27 % reduction in dust particles in London (Hyde Park) by green cover of 2.5 km^2 . Planting of trees and shrubs was recommended as a way to combat dust pollution in Russian cities and there was 2-3 times reduction in dust fall by planting a 8 m wide green belt between the roads and buildings. (Singh & Verma, 2006). Dochinger (1980) examined the ability of plants to abate particulate pollutants and reported a reduction of up to 42 % in overall dust fall by a canopy of coniferous plants in the urban areas of Ohio, U.S. Varshney &

Mitra (1993) assessed the particulate abatement capacity (PAC) of three commonly grown hedge species and found their PAC in the following order –

Duranta plumieri > *Bougainvillea spectabilis* > *Nerium indicum*

They reported that the row of roadside hedges trapped nearly 40 % of SPM, most of which arise from the traffic movement

Phyto-remediation of Indoor Air Pollution
Deteriorating quality of the indoor breathing air by in-house air pollutants is a major concern today as people spend nearly 80 to 90 % of their time in their house or other indoor premises. Studies indicate that pollution levels in the homes are up to 20 times worse than the air outside. The main sources could be wall paints, carpets, furniture & fixtures, gas stoves in kitchen. The pollutants more widely present in the indoor environment are –

1   Carbon monoxide (CO)
2   Nitrogen oxides (NO_x)
3   Formaldehyde
4   Volatile Organic Compounds (VOCs) e.g. toluene, xylene, ethylebenzine & chloropyrifos
5   Undesirable products of burning tobacco & wood

Over 300 indoor plants have been identified that has the wonderful property of absorbing and removing indoor air pollutants such as benzene, formaldehyde, trichloroethylene and all other VOCs. They can also extract particulate matters from the air. (Wolverton & Wolverton, 1993; Wolverton, 1997; Orwell et al. 2004). Plant roots can also absorb pollutants and render them harmless in soils.

In a study sponsored by the National Aeronautics and Space Research Agency (NASA), spider plants (*Chlorphytum elatum*) were placed in a closed chambers with 120 ppm of CO in one and 50 ppm of NO_2 in other. After 24 hours the spider plants removed 96 % of the CO and 99 % of the NO_2. Experiments with pothos plants (*Eppiremnum aureum*) showed that 75 % of the CO was removed after 24 hours. (Wolverton, 1985).

Following broad-leaved indoor plants have been identified to remove formaldehyde & other contaminants including the VOCs from the indoor air-

1   Heart-leaf philodendron (*Philodendron scandens*)

2   Elephant-ear philodendron (*Philodendron domesticum*)
3   Green spider plant (*Chlorphytum elatum*)
4   Golden pothos (*Eppiremnum aureum*)
5   Peperomia (*Peperomia obtusifolia*)
6   Peace lily (*Spathiphyllum clevelandii*)
7   Snake plant (*Sansevieria traifasciata*)
8   Chinese evergreen (*Aglonema modestum*)

GREEN PLANTS : THE POTENTIAL CARBON SINK ON EARTH REDUCING GLOBAL WARMING

Plants on land (forest) or in oceans (phytoplanktons) remove vast amounts of CO_2 from atmosphere. They are natural 'carbon sinks' on earth. Plants lock up the carbon dioxide from the atmosphere as plant material (mainly complex carbohydrates) in the process of photosynthesis for a specified period of time possibly up to 100 years. At present, terrestrial ecosystems (forest and vegetation on land) absorb about 25-30 % of the CO_2 emitted by human activities, thus providing a valuable free environmental service to slow the rate of global warming and climate change. Each year 16 % of atmospheric carbon dioxide is cycled through green plants, and forest contain up to 85 % of all carbon that is bound up in living organisms. Considerable amount of this remains in the ground as soil organic matter (SOM).

$$6 H_2O + 6 CO_2 \text{ (green plants-sunlight} \rightarrow \text{photosynthesis)} \rightarrow C_6H_{12}O_6 + 6 O_2$$

Temperate forests sequester about 2.7 tons of carbon per hectare a year for the first 80 years of their lives. In temperate areas about 400 million hectares (more than the current forested area of the U.S.) of growing forests will be required to sequester 1 billion of the 3-4 billion tons of carbon accumulates in the atmosphere every year. In the tropics, where less carbon is sequestered per hectare (tropical forests are estimated to fix between 1 to 2 tones of CO_2 for each square kilometer of land area every year) locking up 1 billion tons of carbon a year would require about 600 million hectares of growing forest, the equivalent of about 75 % of the area of the Amazon basin. (World Development Report, 1992).

The Novel Concept of Carbon Credits

Governments in Australia and European nations are developing a strategy which require industries and motorists emitting carbon dioxide (CO_2) to buy '*carbon credits*' from organizations/farmers which grow trees or manage existing vegetation. Sequestered or absorbed carbon is locked for long period, in order to provide full offset for the CO_2 emitted by our industries and automobiles. The growth of a typical tree tends to be slow in the early years as the trees establish themselves. Trees grow and sequester carbon fastest when they are about 10 to 20 years old. Study has shown that a 500 MW coal-fired power plant operating over a 35 year lifetime would require a tree plantation program over 1400 km^2, and maintained for several hundred years. Forestry programs are estimated to be able to provide a storage capacity of 1.2 gigaton (Gt) carbon per year. Tree plantations have secondary environmental benefits like improving biodiversity corridors, maintaining watershed hydrology and oxygenation of atmosphere.

ROLE OF ENVIRONMENTAL BIOTECHNOLOGY & PLANT GENETIC ENGINEERING IN DEVELOPMENT OF TRANSGENIC PLANTS TO COMBAT AIR POLLUTION

Genetic engineering has produced some 'wonder plants' to combat air pollution. A number of such plants such as *Arabidopsis umbellate, Pittosporum tobire* and *Raphiolepis umbellate* are now available to work as 'natural sink' for the air pollutants. (Singh & Verma, 2006). Key enzymes and their genes involved in metabolising NO_2 such as 'nitrate reductase' (NR), 'nitrite reductase' (NiR), glutamine synthetase (GS) and those involved in metabolism of SO_2 such as 'sulfite oxidase' and 'sulfate oxidase' have been identified. Transfer and 'over-expression' of these genes may help in development of transgenic NO_2 and SO_2 – philic plants which can take up high amounts of these pollutants from the atmosphere and metabolise them. Takahashi et al, (2001) found that enrichment of genes coding for synthesis of 'nitrate reductase' enzymes improved assimilation of nitrogen dioxides (NO_2) in *Arabidopsis* plants.

Chapter 2

PLANTS COMBATING WATER POLLUTION

Several aquatic plants have been reported for long to detoxify environmental pollutants in the water bodies and the aquatic ecosystems. Remediation of water contaminated even with 'chlorinated alkanes' have been shown with aquatic plants. Plant enzymes and the symbiotic microbes on their roots plays important role in biodegradation of complex organics.

CHEMICAL POLLUTANTS IN
NATURAL WATER SOURCES

Natural water, whether surface or groundwater already carry some impurities due to natural causes specially if the catchment areas are not properly managed. Over the years, and with the explosion of human population the natural water sources (rivers, lakes and streams) in the environment have been contaminated by various human developmental activities.

In the wake of rapid industrial development and pollution, some new chemical and biological contaminants have entered into the natural water pools of earth. Both municipal and industrial wastewater contain a wide variety of chemicals in much higher concentrations than in the natural water. Several of them pass into natural water courses as they cannot be removed from the wastewater by conventional treatment technologies.

The effect a chemical pollutant has in water depends on the nature of the pollutant and on factors such as acidity (pH), temperature, water hardness, presence of organic materials such as algae and weeds, and the oxygen content of the water. Very low concentrations of heavy metals and other chemicals in water can have an enormous impact. The toxicity of certain heavy metals tends to increase as the pH of the water decreases.

The toxicity of chemicals in water depends upon two important factors - '*bioaccumulation*' and '*biomagnification*'. Bioaccumulation is the accumulation of a chemical by an organism to a concentration that exceeds that of the surrounding environment. Glaring example is pesticide aldrin whose concentration in the tissues of experimental snails has been found to be nearly 5000 times the concentration of aldrin in the water in which the snails live (WHO Report, 1996). The same is the fate of DDT and mercury (Hg) in water. Several times of their concentration have been found in fishes.

**Table 1. Maximum Permissible Contaminant Levels
for Drinking Water Prescribed by WHO**

Contaminants Permissible	Level Permitted (in ppm)
1. Lead	0.15
2. Arsenic	0.05
3. Copper	1.3
4. Iron	0.3
5. Zinc	5.0
6. Selenium	0.1
7. Chromium (Total)	0.1
8. Manganese (Desired)	0.5
9. Fluoride	4.0
10. Chloride	250
11. Sulfate	500
12. Nitrate (NO_3)	10
13. pH (Desired)	6.5-8.5
14. Total Solids (Desired)	500
15. Cyanide	0.2
Contaminants Permissible	Level Permitted (in ppm)
16. Alkyl Benzene Sulfonate	0.5
17. Carbon Chloroform Extract	0.2
18. Radium 226 & 228 combined	5 pCi/litre
19. Asbestos	7 million fibers/litre
20. Gross Beta Radiation	4 mrem/year

Source : EPA, U.S.A. (1998).

The more a chemical is persistent, the greater the potential for bioaccumulation. The octanol / water partition coefficient of a chemical is a good indicator of its bioaccumulation potential. The higher the coefficient, the more a chemical will tend to bioaccumulate. Chemicals in the environment can also become increasingly concentrated in the tissues of animals including human beings higher up in the food-chain. This is called biomagnification. Several thousand fishermen died, others suffered severe brain damage after eating fish from the minamata bay contaminated by methyl mercury in Japan in 1950.

Following are some of the important chemical contaminants encountered in the natural water sources which have to be removed completely to provide safe drinking water to society-

Organic and Inorganic Solids in Natural Water

Solids in water consists of organic or inorganic particles. Solids can either be suspended or dissolved. Suspended solids includes all organic and inorganic materials suspended in water. They provide adsorption sites for biological and chemical agents, and give microorganisms protection against chlorine disinfectants. Heavy metals, particularly cadmium (Cd), chromium (Cr), lead (Pb), mercury (Hg), selenium (Se) and arsenic (As) may be adsorbed on the suspended solids (SS) in raw water. Surface water often contain inorganic solids, such as clay, silt, and other soil constituents from erosion. Organic solids like plant fibers and bacteria are also common.

Synthetic Organics in Surface and Groundwater

With the development of gas chromatography and mass spectroscopy, it has become possible to identify hundreds of organic compounds in surface and groundwater. Some are biodegradable, others non-biodegradable. The biodegradable organic matters in water or wastewater is expressed as BOD (biochemical oxygen demand). Higher the BOD, more is the concentration of organics in water / wastewater. There are some natural organic matters (NOM) in water, but the majority are synthetic, common of which are pesticides and industrial solvents.

Major industrial organic solvents identified in groundwater are trichloroethylene (TCE), perchloroethylene (PCE), trichloromethane (TCM), dichloromethane (DCM) and trichloroethane (TCA). Others are synthesized when NOM in water reacts with chlorine during disinfection process.

Organic compounds in the industrial wastewater comprise a broad range of industrial solvents, disinfectants and other pesticides, gasoline, resins, polychlorinated biphenyls (PCBs), dioxin, phenol and chemical reagents. Phenol is an acutely toxic liquid chemical waste in wastewater with a benzene ring (C_6H_5OH). It is mainly used or produced in integrated steel mills, synthetic textile mills, and in resin (plastic) manufacturing. PCBs and dioxins are also the most dangerous chemicals known to man.

NDMA (N-nitrosodimethylamine), a principal ingredient in rocket fuel and MTBE (methyl tertiary butyl ether), a highly soluble gasoline additive are new discoveries in surface and ground water. New, highly complex, organic compound 'acrylonitrile' (produced by the chemical industry for the use in textile industry as a raw material for the manufacture of certain new synthetic fibers) discovered in streams are extremely toxic.

Other organic compounds found in stream waters are ciprofloxacin, amoxicillin, doxycycline, norfloxacin, tetracycline, sulfamethaxazole and trimethoprim – all antibiotics; anthracene, benzopyrene, naphthalene, phenanthrene- all polyaromatic hydrocarbons (PAH); chlordane, coarbaryl, diazinon, dieldrin, chlorpyrifos, methyl parathion, lindane – all pesticides.

Nearly all organic compounds found in water are toxic, carcinogenic (inducing cancer) and mutagenic (causing birth defects). Organic matter in water may become the host for several opportunistic microorganisms. As microbes metabolize organic matters, they consume oxygen and can lead to oxygen depletion and deterioration in natural water quality.

The Chemical Disinfection By-products in Drinking Water

Raw water has to be disinfected before community supply and chlorination is the most common disinfection process. When chlorine is added to water containing natural organic matters a variety of toxic organic compounds are formed. Although present in low concentrations many of them are known or suspected potential human carcinogens. Important among them are trihalomethanes (THMs), haloacetic acids (HAAs), trichlorophenol, and aldehydes.

Ozonation is alternative method of water disinfection by the use of ozone gas (O_3). Ozone also reacts with most organic matter in the water and the byproducts formed are organic peroxides, unsaturated aldehydes, and epoxides, which are of health concerns to humans.

Chlorinated Organic Solvents in Water

Major chlorinated industrial solvents identified in groundwater are carbon tetrachloride (CTC), trichloroethylene (TCE), perchloroethylene (PCE), trichloromethane (TCM), dichloromethane (DCM), dichloroethane (DCA), trichloroethane (TCA) tetrachloroethylene, methyl chloride and vinyl chloride. They are released in their manufacture in the petrochemical industry or in the resulting solvent use in drycleaning industries. These substances have contaminated many deep-lying waters through industrial discharges and dumping into disused wells and leachates. They are highly carcinogenic and mutagenic and must be removed from water.

Refractory Organics (ROs) in Water

Refractory organics are resistant to biodegradation. Constituents of woody plants like tannins and lignic acids, phenols and cellulose are found in natural water systems. Other non-biodegradable constituents in raw water are some polysaccharides with exceptionally strong bonds and benzene with its ringed structure discharged from the petroleum refineries. These organics tend to resist conventional methods of water and wastewater treatment technologies.

Complex Hydrocarbons in Water

Hydrocarbons like petrol, kerosene, oils and lubricants have been found in surface and ground waters. They are discharged from oil refineries, gasworks and industrial effluents and fumes. The polycyclic aromatic hydrocarbons (PAH) such as benzopyrene, benzofluoranthene, benzoperylene, benzofluoranthene and indenopyrene are highly carcinogenic and must be removed from water.

Agricultural Pesticides in Water

Several agricultural pesticides have been found in surface water resulting from the run-off from the agricultural fields. Important pesticides discovered in raw water are aldrin and dieldrin, DDT (all isomers), chlordane (all isomers), heptachlor and hexachloro-epoxy, methoxychlor, dichlorophenoxy-acetic acid, simazin and attrazine They are highly carcinogenic and mutagenic.

Nearly 300 UK water supplies were found to be contaminated with pesticides above the WHO limits. (UNEP Report, 2004).

Explosives Contaminated Groundwater

There are numerous defense disposal sites across the USA with groundwater contaminated by residues of explosives such as TNT (Trinitrotoluene), RDX (Royal Demolition Explosive) and HMX (High Melting Explosives) DNT.

Surfactants (Surface-Active Agents) in Water

Surfactants are large non-biodegradable organic molecules that are slightly soluble in water, and cause foaming in surface waters and water treatment plants. Surfactant tend to collect at the air-water interface and create a very stable foam. Typical example are the synthetic detergent 'alkyl-benzene-sulphonates' (ABS) introduced in 1950.

Volatile Organic Compounds (VOCs) in Water

Many industrial wastewater contain VOCs that may be flammable, toxic, and odorous. Typical example are vinyl chloride, 1,1,1-trichloroethane, trichloroethylene, benzene, ethylbenzene, methylene chloride, and toluene, di-n-butyl phthalate, naphthalene, p-chloro-m-cresol, and phenol; pesticides e.g. DDT, dieldrin, and heptachlor; and the polychlorinated biphenyls (PCBs).

VOCs are of great environmental and health concern because once such chemical is in the vapour state they are much more mobile and also contribute to a general increase in the reactive hydrocarbons in the atmosphere, leading to the formation of 'photochemical smog' and 'tropospheric ozone'. VOCs are also potential carcinogens.

Several VOCs pass through the conventional wastewater treatment plants and contaminate receiving surface waters (rivers and streams). Vinyl chloride, trichloroethane, trichloroethylene, benzene, ethylbenzene, methylene chloride, and toluene have been found in surface waters and groundwater.

Persistent Organic Pollutants (POPs) in Water

Certain chemicals in the water persist in the environment for exceptionally long period of time. They resist photolytic, chemical and biological degradation. They are semi-volatile, highly toxic and transported to the remotest corner of Earth, even up to the Artic.

Typical examples of POPs are high-molecular weight chlorinated aromatic hydrocarbons, pesticides like aldrin, dieldrin, endrin, heptachlor, chlordane, mirex, hexachlorobenzene, indane, and DDT; PCBs, phthalates, poly-chlorinated dioxins and furans, and the toxaphene. DDT can remain in the environment for decades. Dioxins are one of the most stable and dangerous chemical compound ever know to the civilization.

Endocrine Disrupting Chemicals in Water

A wide array of industrial chemicals (some 45) specially the synthetic chemicals and the POPs in water sources can disrupt the genetically based messages through disruption of 'endocrine function' causing reproductive and developmental abnormalities, neurological and immunological problems and cancer.

Inorganic Compounds in Water

Natural water sources have been found to contain several inorganics in the form of nutrients e.g. nitrogen, phosphorus, calcium, iron, potassium, manganese, cobalt, boron and sulfur. Fluorides, chlorides, arsenic, metals including heavy metals, are other important inorganics found in natural water whose concentration is increasing.

Nitrogen (N) and phosphorus (P) are alarmingly increasing in surface waters. Excess amount of phosphorus and nitrogen contribute to 'algal blooms' and *eutrophication* (oxygen depletion and deterioration) of surface waters. Complex inorganic phosphates, such as P_2O_5, at levels as low as 0.5 ppm, can interfere with normal coagulation and sedimentation processes in water-purification plants.

Nitrate in Water

Nitrate (NO_3) pollution of drinking water is a growing concern as human health hazard worldwide with increased use of chemical nitrogen urea (NH_2CONH_2) in modern agriculture, fertilizing lawns and gardens, parks and golf courses, from where they easily run-off into the local water courses or seep into underground aquifers. Intensification of livestock production is also releasing nitrogen rich liquid manure into the environment. Nitrogen is also originating from waste disposal, industrial effluents, including paper & munitions manufacture. Nitrifying bacteria (*Urobacteria, Nitrosomonas & Nitrobacter*) in the soil convert the chemical & other forms of nitrogen into 'nitrate' which is the cause of concern. Being highly soluble nitrate readily leaches through soil and move into groundwater. It is undetectable in water without testing as it is colorless, odorless, and tasteless.

The denitrifying bacteria (*Bacillus, Chromobacter, Pseudomonas, Spirillium* etc.) in soil help in converting the dangerous 'nitrates' to harmless gaseous nitrogen (N_2) but there is associated risk of generation of greenhouse gases 'nitrous oxide' (N_2O) and carbon dioxide (CO_2) in the process. N_2O also plays role in destruction of stratospheric ozone.

In the recent years concern for nitrates in potable water has been growing alarmingly as nitrate is a potential health hazard. WHO recommends 10 mg/L of nitrate-nitrogen in drinking water as the safe limit. Nitrate concentration 10 – 45 mg/L or more is considered to be 'carcinogenic' (due to formation of 'nitrosamines in guts) and causative factor for the 'blue-baby syndrome' (methemoglobinemia) in which the nitrate in gut is converted into 'nitrite', which then combines with blood hemoglobin to form 'methamoglobin' thuis decreasing the ability of the blood to carry the vital oxygen. Nitrate rich water is also reported to affect the human CNS (central nervous system) and CVS (cardiovascular system).

Fluoride in Water

With the rising geological explorations and weathering of fluoride bearing rocks (fluorapatites), excessive fluoride have started entering into groundwater. The aluminum and phosphate fertilizer industries are also contributing significant amount of fluoride into the human environment.

Excessive discharge of groundwater (which is occurring in most part of world) leads to increase in concentration of fluoride in water. The WHO permissible limits of fluoride in water is 0.7 to 1.2 ppm above which it may have toxic effects.

Arsenic (As) in Water Inorganic arsenic of geological origin is found in groundwater in several parts of world including in India and Australia. It is highly persistent in the environment and can pass into human food chain. Fish bio-accumulate arsenic as organic compounds.

In well –oxygenated water and sediments, nearly all arsenic is present in the stable form of arsenate. Some chemical forms of arsenic adhere strongly to clay and organic matter. There is potential for arsenic to be released from water and sediments. Since many arsenic compounds adhere strongly to soils, water percolating down does not usually move arsenic through more than a short distance in soil into the deep groundwater aquifers.

High levels of arsenic in drinking water is proving carcinogenic and also mutagenic (causing birth defects). The WHO established 10 ppb (parts per billion) as a provisional guideline for safe limit of arsenic in water in 1993. Thousands of US water-supply systems exceed the WHO limit of arsenic in water and it is 5 times higher. This is linked with high risk of prostrate cancer.

Sources & State of Heavy Metals in Water Bodies

Developmental activities and their byproducts are the sources of metal contamination of our soils & water bodies. Main sources are –

1   Metalliferous Mining & Smelting : Arsenic (As), cadmium (Cd), lead (Pb) & mercury (Hg);
2   Industries (Metal & Electroplating, Saw Mills etc.) : Chromium (Cr), cobalt (Co), copper (Cu), Zinc (Zn), nickel (Ni), As, Cd & Hg;
3   Atmospheric Deposition from Industries & Automobiles: Uranium (U), As, Cd, Cr, Cu, Pb & Hg;
4   Agricultural activities: Selenium (Se), As, Cd, Cu, Cr, Pb, Zn & U
5   Waste disposal (MSW landfills & wastewater treatment plants): As, Cd, Cr, Cu, Pb, Hg, & Zn

Some heavy metals, such as cobalt (Co), chromium (Cr), copper (Cu), nickel (Ni) & zinc (Zn) are essential and serve as micronutrients for plants like calcium (Ca), potassium (K), magnesium (Mg), manganese (Mn), iron (Fe) and sodium (Na). They are used for redox-processes, as components of various enzymes and for regulation of osmotic pressure in cells. Other metals have no biological role e.g. cadmium (Cd), lead (Pb), mercury (Hg), aluminum (Al), gold (Au) and silver (Ag). They are non-essential and potentially toxic to soil microbes. Some of them e.g. Cd^{2+}, Ag^{2+}, Hg^{2+} tend to bind the SH groups of enzymes and inhibit their activity (Turpeinen, 2002). At high concentrations, both essential and non-essential metals can damage cell membrane, alter enzyme specificity, disrupt cellular function, and even damage the structure of DNA.

PLANTS INVOLVED IN COMBATING WATER POLLUTANTS

Several aquatic plants from green algae (non-seeded plants) to angiosperms (seeded plants) are being identified and experimented to be used in phytoremediation of chemically contaminated (by toxic organics and heavy metals) water and wastewater. The floating hydrophyte water hyacinth (*Eichhornia cressipes*) can remove heavy metals by 20-100 %. Another freshwater free floating species *Ceratophyllum demersum* serve as a 'biofilter of toxic metals'. It can bio-accumulate arsenic (As) with a 20,000 – fold concentration factor. (Weis & Weis, 2004). Another versatile species is *Myriophyllum aquaticum*. It has been successfully tested for phytoremediation of soils contaminated by trinitrotoluene (TNT) as well as trichloroethylene (TCE) and PCP. It can also phytotransform perchlorate. (Susaral et al., 1999).

The brake fern (*Pteris vittata*) is hardy, versatile and fast-growing perennial plant easy to propagate. It is extremely efficient in extracting arsenic (As) from both soil and water and translocating it into its above-ground biomass mainly the fronds. It can reduce the concentration of 200 ppm of arsenic in water to nearly 100-fold (10 ppm) within 24 hours and can hold up to 22 grams of arsenic per kg of plant matter.

Phytoremediation has also been successfully tested for remediation of radionuclides such as uranium (U^{238}) from wastewater in Ohio, U.S and remediation of cesium (Cs^{137}) and strontium (Sr^{90}) from a pond near Chernobyl, Ukraine. The nuclear accidents at Chernobyl in Ukraine (1986) spewed hazardous radioactive materials in the lakes & ponds located around the NPP. Some green algae like *Chara, Nitella* and *Ulothrix* have been identified as performing great phytoremediation functions. If they could be induced to grow in uranium mining effluents, they would provide a simple, long-term solution to remove uranium (U) and other radionuclides from contaminated water. (Prasad, 2006).

The most versatile plant species, both aquatic and terrestrial (grown hydroponically) that has been identified for phytoremediation (of both chemically and radioactively contaminated water) after rigorous laboratory and field experiments are –

Table 2. Aquatic Plants Capable of Removing Chemical Pollutants from Water

Common Aquatic Plants	Mechanism of Phytoremediation
Azolla filiculoides (Water fern)	Metal hyper-accumulation
Apium graveolens (Celery)	Removes sulphonated anthraquinones from textile wastewater
Bacopa monnieri (Water hyssop)	Metal accumulation
Carex gracillus	Degrades trinitrotoluene (TNT) & uptake uranium (U)
Canna flaccida (Water lily)	Removal of heavy metals and nitrates
Ceratophyllum demersum (Coontail)	Degrades complex organics, remove radionuclides and hyper-accumulates arsenic (As)
Eichhornia cressipes (Water hyacinth)	Metal hyper-accumulation, removal of radionuclides & removal of nitrates
Eirophorum angustifolium	Phyto-stabilization of submerged metal rich mine tailings
Elodea canadenesis	Removal of nitrates from water
Eleocharis tuberosa (Water chestnut)	Transformation of TNT
Glyceria fluitans	Phyto-stablization of mine tailings & treatment of acid mine drainage
Hydrilla verticillata (Pond scumb)	TNT transformation and metal accumulation
Hydrocotyle umbellate (Pennywort)	Accumulation of toxic metals
Ipomea aquatica (Water spinach)	Metal accumulation
Lemna minor (Duckweeds)	Heavy metal accumulation, concentration of technetium-99, nitrate removal
Myriophyllum aquaticum (Parrot feather)	Degradation of perchlorate & organophosphates
	Transformation of halogenated organics (TCE, PCP)

Table 2. (Continued).

Common Aquatic Plants	Mechanism of Phytoremediation
	Halocarbon metabolism
Nymphaea violacea (Water lily)	Uptake of radionuclides – uranium & thorium
Nelumbo nucifera (Indian lotus)	TNT transformation
Potamogeton Spp. (Pond weed) Polygonum punctatum	Degradation of explosives & heavy metals uptake Uptake of readionuclide cesium-137 (Cs137)
Rumex hydrolapatum	Removes sulphonated anthraquinones from textile wastewater
Salvinia rotundifolia (Floating moss)	TNT transformation
Typha Spp. (Cattails)	Degrades TNT and perchlorate, removes nitrates
Tamarix Spp. (Salt cedar)	Hydraulic control of arsenic (As) in water
Vallisneria americana (Tape grass)	Transformation of trichloroethane (TCE) & metal uptake
Wolffia Spp. (Duckweeds) Zizania aquatica (Wild rice)	Absorption of organic & inorganic pollutants Uptake of radionuclide iodine -129 (I129)

Source: Prasad (2006) & also from other authors cited in references.

Table 3. Terrestrial Plants Adapted to Grow Hydroponically for Removal of Pollutants in Water

Important Species	Phytoremediation Functions
Brassica juncea (Indian mustard)	Heavy metal removal
Hellianthus annus (Sunflower)	Removal of radionuclides & heavy metals
Vetiveria zizionides (Vetiver grass)	Heavy metal removal (Hyper-accumulator)

Table 4. Aquatic Plants Used for Bio-monitoring & Bio-accumulation of Heavy Metals from Water

Important Aquatic Plant Species	Metals
Azolla Spp. (Water fern)	Cr, Ni, Zn, Fe, Cu, Cd & Pb
Bacopa monnieri (Water hyssop)	Hg, Cr, Cu & Cd
Carex Spp.	Cu, Pb, Zn, Co, Ni, Cr, Cd, Fe, Mn & Mo
Ceratophyllum demersum (Coontail)	As, Cd, Cu, Cr, Pb, Hg, Fe, Mn, Zn, Ni & Co
Eichhornia cressipes (Water	As, Cd, Co, Cr, Cu, Al, Ni, Pb, Zn, Hg, P, Pt,

hyacinth)	Pd, Os, Ru, Ir & Rh
Elodea candensis	Cu, Cd, Pb, Zn, Cr & Ni
Hydrilla verticillata	Hg, Fe, Ni & Pb
Lemna minor (Duckweeds)	Mn, Pb, Ba, B, Cd, Cu, Cr, Ni, Se, Zn & Fe
Ludwigia natans	Hg & Methyl Mercury (Hg)
Myriophyllum Spp. (Parrot feather)	Cd, Cu, Zn, Pb, Fe, Hg, Ni & Cr
Nymphaea Spp. (Water lily)	Ni, Cr, Co, Zn, Mn, Pb, Cd, Cu, Hg, & Fe
Potamogeton Spp. (Pond weed)	Ni, Cr, Co, Zn, Mn, Pb, Cd, Cu, Hg, As, Se & Fe
Pistia stratoites (Water lettuce)	Cu, Al, Cr, P & Hg
Ranunculus aquatilis	Mn, Pb, Cd, Fe & Pb
Salvinia Spp. (Water fern)	Cd, Fe, Pb, Cr & Mn
Scapania uliginosa	B, Ba, Cd, Co, Cr, Cu, Li, Mn, Mo, Ni, Pb, Sr & V
Typha latifolia (Cattails)	Ni, Cr, Co, Zn, Mn, Pb, Cd, Cu, Hg & Fe
Vallisneria americana (Tape grass)	Cd, Cr, Cu, Ni, Pb & Zn

Source : Prasad (2006).

MECHANISM OF REMOVAL OF POLLUTANTS FROM WATER BY PLANTS

Phytoremediation technology in aquatic medium works mainly through –

Rhizo-Filtration

It is based on a combination of principle of phytoextraction and phytostabilization specially suited to remove metals and radionuclides from polluted water. Contaminants are absorbed and concentrated by plant roots, then precipitated as their carbonates and phosphates (Salt *et al.* 1995).

Rhizofiltration also works in the efficient removal of radionuclides and toxic organics such as tetrachloroethane, trichloroethylene, metachlor, atrazine, nitrotoluenes, anilines, dioxins and various petroleum hydrocarbons (Rice *et al.* 1997).

Phyto-Volatilization

Plants absorb and transpire the impurities from soil and water through their aerial organs. Some contaminants like selenium (Se), mercury (Hg) and volatile organic compounds (VOCs), can be released through the leaves into the atmosphere. (Cunningham and Ow, 1996).

Phyto-stimulation : Plant – Assisted Microbial Degradation (Rhizosphere Biodegradation)

Many organic compounds are degraded by microorganisms located in the rhizospheres (on the roots) of aquatic plants. Certain plant roots release substances that are nutrients for microorganisms, bacteria and fungi, that provide favorable habitats for soil microbes to act (Cunningham and Ow, 1996). A typical microbial population present in the rhizosphere, per gram of air-dried soil comprises – bacteria 5 x 10^6, actinomycetes 9 x 10^5, and fungi 2 x 10^3 (Schnoor et al.,1995).

A large variety of exudates are secreted from roots in the form of sugars, amino acids, essential vitamins & enzymes. Root exudates may also include acetates, esters and benzene derivatives. Enzymes are also present in the rhizosphere and this may act as substrates for the microbial population. This results in increased biological activity of the microbes in the area immediately surrounding the root zone (rhizosphere).

By encouraging a microbiologically active rhizosphere, the plants facilitate accelerated digestion (biodegradation) of wide variety of organic contaminants in the upper soil layers and / or wastewater / polluted water (Anderson et al. 1993). The water hyacinths (Eichhornia crassipes) works on the same biological principle. It harbours several microbes in its root zone which perform the task of biodegradation of heavy metals in polluted water and also helps in absorption and adsorption of chemical impurities. Interestingly, gram-negative bacteria appear to have some important metabolic capabilities for degrading xenobiotic chemicals not found in gram-positive bacteria.

Enzymes secreted by plant roots or the microbial community in the rhizosphere comprise esterases different oxido-reductases (phenoloxidases and peroxidases). Plant peroxidases are exuded by some members of Fabacea, Graminae and Solanaceae. White radish (Raphanus sativus) and horse radish (Armoracia rusticana) secrete 'peroxidase' while the parrot feather (Myriophyllum aquaticum) and the aquatic green algae Nitella & Chara secrete 'laccase'. (Dias et al., 2006). Aquatic plants grown in 'constructed wetlands' (described below in this chapter) provides greater surface area for microbial association and growth with enhanced phytoremediation activities.

FATE OF POLLUTANTS ONCE ACCUMULATED INTO AQUATIC PLANTS

Phyto-Degradation

Certain plant species breakdown the contaminants after absorbing them. This they do through enzyme-catalyzed metabolic process within their root or shoot cells. Others breakdown the contaminants in the substrate itself by secreting enzymes and chemical compounds. The enzymes secreted are usually dehydrogenases, oxygenases and reductases. The biodegraded constituents are converted into insoluble and inert materials that are stored in the lignin or released as exudates (Watanabe, 1997). Some plants biodegrade contaminants with the aid of microbes which live in symbiotic association on their roots;

Table 5. Plant Enzymes Implicated in Phytodegradation & Phytotransformation of Organic & Inorganic Compounds

Enzymes	Contaminants Degraded / Transformed into Less Toxic Forms
1). Phosphatase	Organophosphates
2). Aromic Dehalogenase	Chlorinated aromatic compounds (e.g. DDT, PCBs etc.)
3). O-demythlase	Metaoalchor, Alachlor
4). Cytochrome 450, Peroxidases & Peroxygenases	PCBs
5). Glutathione s-transferase, carbo-oxylesterases o-glucosyltransferases, o-malonyltransferases	Xenobiotics
6). β-cyanoanaline synthase	Cyanide

Sources : (Sandermann, 1994; Macek et al., 2000; Prasad, 2006).

Phyto-Transformation

Several inorganic and organic contaminants once absorbed inside the root, may become biochemically bound to cellular tissues (biotransformed), in forms that are biologically inert or less active (Watanabe, 1997). In many cases, plants have the ability to metabolize organic pollutants by phytotransformation and conjugation reactions followed by compartmental-izing products in their tissues.

REMOVAL OF NITRATES FROM GROUNDWATER

Three aquatic plants *Cladophora* spp. (green algae), *Elodea Canadensis* & *Scirpus pungens* have been shown to effectively remove nitrates (NO_3) from nutrient enriched water bodies. Plants such as bulrush, arrowhead, cattail, sweet flag, water hyacinth, duck weeds, bamboo and poplar have been shown to clean the water polluted with nitrates and make it safe for human and animal use. (Dwivedi, 2006).

A novel biotechnological method for nitrate removal from drinking water has been developed in Germany where 'nitrate' (NO_3) is converted into 'nitrogen' (N_2) gas with the aid of plant enzymes. Nitrate containing groundwater is passed through a column where three enzymes – nitrate reductase, nitrite reductase and nitrous oxide reductase are co-immobilized along with electron-carrying dye which are energized by electric current. This provides the reducing power to drive the conversion of nitrate to nitrogen gas and water. Prototype columns for field tests have been made to process & treat 500 L of nitrate contaminated groundwater per minute. Nitrates are completely removed in a single pass and no contaminant residues are left in water. (Mobetec GmbH, German Patent Application, 1990) (Dwivedi, 2006).

The technology is also being used to cleanup huge nitrates wastes containing organic nitro-compounds accumulated in defense laboratories involved in explosive manufacture such as TNT (Trinitrotoluene), RDX (Royal Demolition Explosive) and HMX (High Melting Explosives) in the U.S.

USE OF DRY BIOMASS OF AQUATIC PLANTS FOR TREATMENT OF TEXTILE EFFLUENTS

Dry biomass of aquatic plants like the water hyacinths (*Eichornia crassipes*) and the giant duck-weeds (*Spirodela polyrrhiza*) has been shown to remove chemical dyes and heavy metals from the textile effluents. Dried roots of water hyacinths removed 'methylene blue' and basic blue dyes efficiently

(Low et al., 1995). Similarly, the dried giant duck weeds can remove methylene blue at broad pH range of 3 – 11. (Warunasantigul et al., 2003).

They have significantly lower operational costs than the convention treatment by 'activated carbons' (activated sludge methods) and they also absorb and remove the heavy metals from the effluents. Moreover, about 90 % of the reactive dyes persists after being subjected to activated sludge treatment.

SOME WONDER AQUATIC PLANTS FOR COMBATING WATER POLLUTION : POTENTIAL FOR COMMERCIALIZATION

The Water Hyacinths (*Eichhornia cressipes*)

Water hyacinths are green floating hydrophytes with large flat leaves and a long spongy petiole partly submerged in water and anchoring in the soil with fibrous roots. It is globally discredited as noxious aquatic weed of ponds, lakes and wetlands but they are endowed with some unique environmental properties for cleaning the polluted water bodies.

Water hyacinths harbor large number of microorganisms in symbiotic relationships on their roots which feed off the minerals and organic chemicals (contaminants) from the effluents. The microbe digest wastes and pollutants and produce sugars and amino acids which are absorbed by the host plant roots. In turn the host plant supply oxygen and nutrients to the microbes for rapid biochemical action and also restore oxygen in the vicinity of the pond and regulate carbon dioxide levels by photosynthesis.

Water hyacinth can remove heavy metals by 20-100 %. In just 24 hours the weed can extract more than 75 % of lead (Pb) from contaminated water. It also absorbs cadmium (Ca), nickel (Ni), chromium (Cr), zinc (Zn), copper (Cu), iron (Fe) and pesticides and several toxic chemicals from the sewage. In just 7 days of exposure it can lower BOD by 97% and remove over 90% of nitrates and phosphates. It can also remove radioactive substances. Water hyacinth could take up significant amount of Sr^{90} rapidly (Jayaraman & Prabhakar, 1982). It also removes algae and fecal bacteria, the suspended matters and removes odor causing compounds. In fact, the weed fosters the growth of a zooplankton *Daphnia* that feed on the harmful pathogens.

Table 6. Treatment of Municipal Sewage by
Water Hyacinth (In mg / L)

Parameters Studied	Raw Municipal Sewage	Treated Sewage	
		After 15 days	After 23 days
1. TSS	320	135	70.5
2. BOD	310	70	9.6
3. Total Nitrogen	65.1	40.1	3.4
4. Total Phosphorus	10.6	4.6	1.1
5. Cadmium	1.6	1.0	0.3
6. Coliform Bacteria	140 x104	21x104	1x104

Source : Tripathi & Shukla (1991), In WWF (India) Report, 1991.

Success Stories

(1).The Boston University, the University of Florida and the Texas State Department of Health Resources in the US is treating sewage with water hyacinths (*Eichhornia*) on experimental basis and has obtained excellent results. The City of San Deigo in California, US, is spending $ 3.5 millions in cultivating water hyacinth for wastewater treatment. The plants are grown hydroponically in a filter of rocks through flows the waste water.(Report of the US Magazine SPAN in India, 1991).(UNEP Report, 1991).

(2). In India, the Central Leather Research Institute, Madras, has successfully used water hyacinths to clean tannery effluents. Water hyacinth has also been used to augment sewage treatment in Calcutta's Salt Lake Swamp. The entire sewage of the city estimated to be around 680 million liters per day is discharged into these lakes. The treated sewage is then utilized for fish farming and irrigation of rice and vegetables in the adjoining farms. (Report of the WWF (India) 1991).

(3). Several exploratory research, pilot plant and full scale studies have been conducted in India at the University of Roorkee, Banaras Hindu University (BHU), Varanasi and at the University of Rajasthan, Jaipur (by author), with very successful results.

Environmental Cost- Benefit Analysis of
Phytoremediation by Water Hyacinth

(1). It saves tremendous amount of energy (electricity) and prevent the greenhouse gas emitted and the chemicals used in the conventional mechanical treatment of wastewater.

(2). The weed biomass of water hyacinths generated as by-product is used to produce biogas and biofertiliser. The ash content of water hyacinth contains 30 % potash and 13 % lime making it an excellent fertilizer.

(3). Each kilogram of water hyacinth by dry weight yields about 370 litres of biogas with average methane content between 69-91 %. The calorific value of the gas when used as fuel is about 580 Btu/ft^3

(4). One hectare of sewage pond in U.S. were found to yield between 20-40 tonnes of water hyacinth per day producing approximately 70,000 cum of · biogas.

(5). Water hyacinth has found new uses in the paper and board industries and can be moulded into cement boards and used as a substitute for the dangerous asbestos.

The Duckweeds (Lemna spp.)

Duckweeds (*Lemna* Spp.) are tiny floating aquatic plants common in lakes, ponds and the freshwater wetlands. They form a green mat over the water surface. They are also endowed with unique environmental restoration properties. They can 'absorb' and 'adsorb' all the dissolved gases and substances, including the heavy metals, from the wastewater. Within 2 to 3 weeks the quality of wastewater improves significantly in terms of BOD and DO values, heavy metals and suspended solids and becomes useful for irrigation, industrial uses and aquaculture.

Duckweeds purifies the wastewater rich in phosphorus, nitrate and potassium until the water is crystal clear with phosphorus and nitrogen contents coming down to 0.5 mg/litre within 20 days. (UNEP Report, 1991).It prevents unwanted algal growth and 'eutrophication' by cutting sunlight below.

In an investigation of toxicity removal and fate of phenol in water treated by *Lemna gibba*, almost 90 % of the applied phenol disappeared from the contaminated water over a 16 days growth period. (Barber *et al.*)

Success Stories

(1). In Bangladesh, the duckweeds are being used under an UNDP sponsored project to treat and purify sewage and the treated sewage ponds are used for fish farming. The duckweed biomass are periodically harvested and used to make cattle feeds.(Shane Cave, UNEP Report, 1991).

2). An American engineer in California, Viet Ngo, has successfully used *Lemna* to clean sewage and wastewater and to convert then into useful material.(UNEP Report, 1991).

Environmental Cost- Benefit Analysis of
Phytoremediation by Duckweeds

(1). Can save tremendous amount of energy (electricity) and prevent greenhouse gas emitted and chemicals used in the conventional mechanical treatment of wastewater.

(2). The weed biomass of duckweeds generated as by-product is used to produce cattle feed and also nutritive feed material for pisiculture (fish farming).

SOME WONDER TERRESTRIAL PLANTS CAPABLE OF GROWING ON WATER FOR COMBATING WATER POLLUTION : POTENTIAL FOR COMMERCIALIZATION

Terrestrial plants like vetiver (*Vetiveria zizanioides*), sunflower (*Helianthus annus*) and Indian mustard (*Brassica junceae*) have large root systems and greater biomass and are especially suitable for growing hydroponically. Species that do not readily transfer contaminants from the roots to stem are preferred, since the accumulated metals and radionuclides can be removed by simply harvesting the roots.

The Indian Mustard (*Brassica junceae*)

The Indian mustard plant has large root system with good biomass and can also be grown hydroponically. It has gained much significance as 'phytoremediator' for environmental cleanup of both chemically & radiologically contaminated soil & water bodies. The Indian mustard concentrate toxic heavy metals (Pb, Cu and Ni) from hydroponics solution to a level up to several percent of their dried shoot biomass (Sriprang, 2006). Several superior 'transgenic species' of *B. juncea* have been produced for improved phytoremediation function.

Success Story

In a study, the Indian mustard grown hydroponically was capable of removing lead (Pb) from aqueous solutions in the range of 4 to 500 mg/L (Dushenkov, et al. 1995). It could effectively remove other metals like Cd, Cr, Cu, Ni and Zn from aqueous medium. (Eapen et al., 2006).

Sunflower (*Helianthus anus*)

Sunflower is a terrestrial plant which can be grown hydroponically in ponds and constructed wetlands for phytoremediation of polluted water. It is reported to absorb radionucleides from water if grown hydroponically. Fortunately, the radioactive materials were bio-accumulated in the roots, shoots and the leaves which was harvested to remove the contaminants and landfilled. The flowers and seeds did not contain any contaminants and hence could be used economically.

Success Stories

1). The technology was successfully employed by Phytotech Inc. (USA) using hydroponically grown sunflower plant (*Helianthus anus*) to remove radionuclides from uranium contaminated water at a DOE pilot project in Ohio, US. It was shown that concentrations of cesium (Cs), strontium (Sr) and uranium (U) from water were significantly reduced within few hours.

2). A phytoremediation study using sunflower plants (*Helianthus anus*) was conducted in the radionuclide contaminated ponds in the vicinity of Chernobyl, Ukraine after the nuclear accident in 1986. It was shown that the sunflower plants grown hydroponically in the pond could take up 90 % of the cesium-137 (Cs^{137}) (from 80 Bq/L of Cs^{137}) in just 12 days. It was estimated that 55 kg of dry sunflower biomass could remove the entire radioactivity from the pond in Chernobyl having 9.2 x 10^6 Bq cesium-137 (Cs^{137}). (Dushenkov, et al. 1999).

3). Phytotech Inc. (USA) and the International Institute of Cell Biology at Kieve, Ukraine conducted study on phytoremediation potential of hydroponically grown sunflower (*Helianthus anus*) and found that they could also effectively remove strontium-90 (Sr^{90}) from ponds in Chernobyl with bioaccumulation concentration of 600 µg/L for both roots and shoots. Study revealed that hydroponically grown sunflower plants reduced Sr^{90} concentrations from 200 µg/L to 35 µg/L within 48 hours and it was further reduced to 1 µg/L.(Dushenkov, 1999).

4). Phytotech Inc. also used sunflower to remove uranium (U) from the contaminated ponds near Chernobyl after the nuclear disaster. Uranium concentration was reduced 10-fold within an hour. Sunflower roots concentrated uranium (U) from solution by up to 10,000 fold. (Eapen, et al., 2006).

Environmental Cost Benefit Analysis of
Phytoremediation by Hydroponically Grown Sunflower

The estimated cost of removal of radionuclides from contaminated water by sunflower ranged from about US $ 2 – $ 6 per thousand gallons of water (adjusting the value of the 'sunflower oil' which is edible oil free of contaminants). It is calculated to be US $ 80 per thousand gallons by physical process. (Terry, 2003).

The Vetiver Grass (*Vetiveria zizaniodes*)

Vetiver has finely structured network of 'deep and spongy root system' often reaching 3- 4 meters in the very first year of growth and can be grown hydroponically. It can tolerate very high acidity and alkalinity conditions (pH from 3.0 to 10.5); very high levels of heavy metals Al, Mn, Mg, As, Cd, Cr, Ni, Cu, Pb, Hg, Se, Zn and the herbicides and pesticides in soils & water. Vetiver can withstand prolonged submergence in water, it also behaves as a wetland plant. It can efficiently absorb dissolved nitrogen (N), phosphorus (P), mercury (Hg), cadmium (Cd), lead (Pb) and all other heavy metals from the polluted streams, ponds and lakes and its efficiency increase with age. Works done in China have confirmed that vetiver can effectively remove dissolved nutrients, specially the N and P from wastewater and reduce the growth of blue green algae (which cause eutrophication) within two days under experimental conditions. Phosphorus (P) is removed up to 99 % after 3 weeks and nitrogen (N) 74 % after 5 weeks. (Zheng, et al., 1998). Works done in Thailand showed that vetiver can also effectively remove substantial quantities of cadmium (Cd), mercury (Hg), chromium (Cr), arsenic (As) and lead (Pb) from municipal wastewater.

Vetiver can easily thrive in wetlands and can be used in the 'constructed wetlands' for removal of nitrogen (N) and phosphorus (P) and heavy metals from the polluted storm water, municipal and industrial wastewater, and effluents from abattoirs, feedlots, piggeries and other intensive livestock industries. An innovative idea is to grow vetiver hydroponically on floating platforms which could be moved from one place to the other, and to the worst affected parts of the lakes and ponds. The advantage of the platform technology is that the top portions of the grass can be harvested easily for stock feed or mulch and the roots can also be removed for oil production.

THE CONSTRUCTED WETLANDS TECHNOLOGY USING AQUATIC PLANTS TO TREAT MUNICIPAL & INDUSTRIAL WASTEWATER FOR REUSE

Based on the understandings of 'chemical breakdown' and 'nutrient removal' properties of aquatic organisms (plants, animals & microbes) in the natural wetland systems, ecologists and environmental biotechnologists are advocating to construct 'artificial wetlands' and utilize them for the treatment of municipal and industrial wastewater, urban stormwater runoff, agricultural wastewater runoff, acid mine drainage and leachates from metal mines and waste landfills.

Over the past 20 years the constructed wetland technology using aquatic plant species and also terrestrial plants with massive root systems capable of growing hydroponically on floating platforms in polluted waters such as the vetiver grass has been widely accepted in Europe and America as a low-cost, environmentally sustainable option to the conventional wastewater treatment facilities which is based on the use of chemicals and high input of energy with consequent emission of greenhouse gases. (Greenway, 2004 & 2006).

Wetland systems treats waste & polluted water by physical, chemical and biotic processes, in a close association of appropriated plants, microorganisms, macro-organisms (vertebrates & invertebrates) and substrates. Macrophytes (rooted emergent plants) enhance physical filtration, prevent clogging in vertical flow systems, mediate oxygen transfer to the rhizosphere and favour microbial colonization. In sub-surface systems, there is an oxygen gradient, with high partial pressures near the plant roots, to be replaced progressively by anaerobic and anoxic environments. The mixture of aerobic, anoxic and anaerobic zones stimulates different microbial communities that can degrade even complex organic pollutants like 'azo dyes' almost to mineralization.

Selection of Aquatic Plant Species for the Constructed Wetlands

Relatively few plants can thrive in high-nutrient, high-BOD, high-sediment waters of constructed wetlands. Among those plants are cattails *(Typha angustifolia & Typha latifolia)*, reed grass *(Phragmites australis)*, the bulrushes *(Schoenoplectus* Spp.) and they have been extensively used in the constructed wetlands to treat even toxic wastewaters. Cattails can degrade explosives including the TNT and the perchlorates. In Australia *Phragmites* has been used extensively. Their root / rhizome systems extend vertically deeper into the sediment and therefore facilitate a large aerobic microenvironment for the biochemical process to occur rapidly. It degrades trinitrotoluene (TNT) and volatalize methyl iodide. Recently *Baumea articulata, Carex fasicularis, Phylidrum languinosum* and *Schoenoplectus mucronta* have been experimented with.

Greenway (2004) has identified 66 native species of macrophytes (8 free-floating, 8 creepers, 5 submerged, 5 floating-leaved attached, and 40 emergent species) growing in surface-flow constructed wetlands receiving secondary-treated effluents in Queensland, Australia. Many species were found growing in effluent containing up to 20 mg /L of $NH_4 - N$, 16 mg/L of NO_x-N, and 9 mg/L of PO_4-P.

Animal Communities Inhabiting the Constructed Wetlands

After the establishment of plant communities, a wide variety of invertebrate and vertebrate animal species come to inhabit the wetlands in a natural process of colonization (or some of them may have to be introduced like plants). Invertebrates include the micro-crustaceans (copepods, ostracods, claderans) rotifers and nematodes. The macro-invertebrates include the annelids (earthworms), molluscs (clams and pond snails), crustaceans (shrimps and crayfish), and insects (dragonflies, damselflies, water beetles, water boatman etc.). Invertebrate species are critical in providing the 'food base' for the vertebrate life of the constructed wetlands which includes the fishes, amphibians (frogs & tadpoles) and reptiles, water-birds and mammals.

Aquatic animals play a significant role as the GRAZERS (herbivores) and PREDATORS (insectivores / carnivores). Grazers consume green plants from large emergent reeds to minute phytoplankton while the predators consume smaller animals to minute zooplanktons. The most important open-water grazers are zooplankton, including rotifers (Rotifera). Grazers often eat the periphyton mats coating water plants. Direct grazing on plants is rare. Water fleas such as *Daphnia* graze significant numbers of bacteria and phytoplankton, and in turn are prey for invertebrate and vertebrate consumers.

Tadpoles feed on detritus and periphyton in the littoral zone. The vertebrate species especially the insectivorous frogs and reptiles play a crucial role in the biological control of the 'mosquitoes' through predator-prey interaction. Some invertebrate like the dragonflies also eat mosquitoes and serve as food for the larger birds and insectivores.

Microbial Communities Inhabiting the Constructed Wetlands

They include photosynthesizing (autotrophic) unicellular and filamentous green algae and blue-green algae (cyanobacteria), heterotrophic (bacteria, fungi and protozoa) and chemo-lithotrophic microorganisms such as the nitrifying bacteria (*Nitrosomonas* Spp.) and the denitrifying bacteria (*Nitrobacter* Spp.). Microorganisms occur in the water column as 'plankton' and attached to surfaces of submerged plants as 'epiphytes' and 'biofilms'. Microbes also occur in the sediments as 'microbenthos'. Anaerobic bacteria occur in low oxygen environments in the sediments.

Certain microbes help to maintain aerobic conditions in the water. These are the autotrophic microorganisms that infuse oxygen into the water by photosynthesis and also absorb nutrients. Among the microbes the nitrifying and the denitrifying bacteria plays the critical roles in improving the water quality. Nitrification is particularly important in the constructed wetlands because the nitrates (NO_3- N) made available can be used by the aquatic plants for growth. Microbes also remove inorganic phosphate from the water column or sediment porewater and convert this to organic microbial biomass.

The Role of Plants, Animals and the Microbes for Removal of Pollutants in Constructed Wetlands

Constructed wetland provides an array of physical, chemical and biological processes to facilitate the removal, recycling, transformation or immobilization of sediments, metals and nutrients. Most of these processes are facilitated by wetland plants, associated biofilms and the microorganisms. Wetlands plants and microbes collectively work to efficiently break down the toxic chemicals in the polluted water, bio-accumulate and bio-transform them.

Aquatic plants grown in 'constructed wetlands' provides greater surface area for microbial association and growth with enhanced phytoremediation activities. Most wetland plants in the solid matrix, can establish associations with bacteria and /or fungi. Plants plays minor role as compared to the microbes.

Microorganisms play critical role in the operation of constructed wetlands. They help reduce BOD, remove and recycle nutrients, heavy metals, hydrocarbons, some pesticides and herbicides from the constructed wetlands by breaking them down in the system. They degrade biodegradable organic materials which descends to the bottom releasing the nutrients contained in them. Microbes also help transform the toxic chemicals into 'non-toxic' forms which is then available to the plants to take up as nutrient material. Heterotrophic microbes feed on the detritus (dead organic matter) and decompose them to release the locked nitrogen and phosphorus. Bacteria feed on the animal matter while the fungi on plant matter.

In the operation of the wetland systems there is continuous interactions between the biotic and the abiotic components and within the biotic components involving the two processes – physical and biogeochemical. In these processes the nutrients from the water (carbon, nitrogen, phosphorus, sulfur and other materials) are either accumulated or removed, recycled or finally stored in the sediments, thereby improving the quality of water column above. Plants and microorganisms remove and recycle nutrients either from the water column or the sediments and incorporate them in their tissues. Most important, bacteria and plants in wetlands can transform inflowing ammonia, nitrate, and phosphate into organic forms, which are later released downstream as detritus. The sediments, biotic components and the detritus are the major storehouse of nutrients.

Rooted macrophytes remove nutrients directly from the sediments, whereas the floating plants remove nutrients from the water column. Submerged macrophytes such as *Potamogeton* and *Ceratophyllum* usually absorb dissolved inorganic nutrients directly from the water column through their leaf surfaces. *Triglochin procera* is also an effective nutrient remover. The floating hydrophytes duckweeds (*Lemna* Spp.) and water hyacinths *(Eichhornia crassipes)* have also been found to be highly effective nutrient and metal remover. Wastewater treatment by duckweeds and the water hyacinth are becoming a preferred technology in India and several other developed and developing countries including the U.S.

Removal of Heavy Metals

Metals are often easily sequestered by the wetland soils or the biota or both. Wetland plants along with the microbes work to break down the sediments and chemicals in wastewater and absorb heavy metals from them. Metal oxidizing bacteria are in the aerobic zones of the wetland. Soil microbes can accumulate metals in tissues in concentrations up to 50 times higher than the surrounding soil. Microorganisms do not actually degrade inorganic metals, but changes their oxidation state. This can lead to an increase in solubility (and subsequent removal by leaching), or precipitation and reduction in bioavailability.

The principal mechanism controlling micro-remediation of inorganic metals are oxidation, reduction, methylation, demethylation, metal-organic complexion, and ligand degradation. To date arsenic (As), chromium (Cr), mercury (Hg) and selenium (Se) have responded well to microbial removal. (Greenway, 2004 & 2006).

Some wetland plants have been found to accumulate heavy metals in their tissues at 100,000 times the concentration in the surrounding water. The submerged aquatic macrophytes have very thin cuticle and therefore, readily take up metals from contaminated water through entire body surface. Further, they redistribute metals from sediments to water and finally take up in the plant tissues and hence maintain circulation. Benthic rooted macrophytes (both submerged & emergent) plays an important role in metal bioavailability from sediments through rhizospehere exchanges and other career chelates. They readily take up metals in their reduced forms from sediments and oxidize them in the plant tissues making them immobile and hence bioconcentrate them to a great extent.

Some wetland plants have been found with inherent metal tolerance and bioaccumulation. Important among them are water hyacinth (*Eichhornia crassipes*), cattail (*Typha latifolia)* and reeds *(Phragmites australis)* and *Glyceria fluitans.* They can be easily established on 'submerged mine tailings'. (Prasad, 2006). They have been used to treat effluents from mining areas that contain high concentrations of heavy metals such as cadmium (Cd), zinc (Zn), mercury (Hg), nickel (Ni), copper (Cu), and vanadium (Va). They transform the toxic heavy metals to non-toxic forms and incorporate them into their body tissues. Metals are mostly stored in the roots and rhizomes of emergent plant species which can be harvested to remove the metals. Metals are removed by adsorption, precipitation, filtration and sedimentation.

Duckweed (*Lemna minor*) and water hyacinth (*Eichhornia crassipes*) have been found to be particularly very effective in removing heavy metals from the contaminated water and render them 'biologically unavailable' to other organisms. As discussed above water hyacinth can remove heavy metals by 20-100 %. In just 24 hours the weed can extract more than 75 % of lead from contaminated water. It also absorbs cadmium, nickel, chromium, zinc, copper, iron and pesticides and several toxic chemicals from the sewage. (Tripathi & Shukla, 1991; Sinha & Sinha, 2000; Sinha et al., 2006).

Removal of Hydrocarbons

Microorganisms break down the complex hydrocarbons in the wastewater by using the three general mechanisms- aerobic and anaerobic respiration and fermentation. Sediments in the wetlands are mostly anaerobic and provide conditions for breaking down complex organics and hydrocarbons (greases, fats, solvents and fuels) and also sequester metals, reducing their bioavailability. The complex 'hydrocarbons' are converted into simpler molecules of 'carbon dioxide' and 'methane' by anaerobic microbes. Anaerobic microbes can degrade the halogens (reductive dehalogenation) and nitrosamine, reduction of epoxides to olefins, reduction of nitro groups and ring fission of aromatic structures. (Nicholas, 1996; Eweis et al., 1998).

Removal of Pesticides

Several aquatic microbes have been identified in constructed wetlands which can break down the chlorinated substances like the herbicides and pesticides. *Flavaobacterium* and *Pseudomonas* can destroy the organophosphates, the fungus *Zylerion xylestrix* can destroy the group of pesticides and herbicides aldrin, dieldrin, parathion and malathion. Majority of the organochlorines appears to be biotransformed, forming cunjugates with the soil humic matter in the wetlands. (Nicholas, 1996; Eweis et al., 1998).

Removal of Pathogens

The biggest challenge in treatment of wastewater is not only the toxic chemicals and high nutrient content but also the sewage-born PATHOGENS. Pathogenic microbes generally pass unaffected through the conventional wastewater treatment plants. Disinfection by chlorination is proving to be a curse in disguise as it results into formation of unwanted by-products the 'trihalomethanes' which are carcinogenic.

Constructed wetlands, however, provide suitable conditions for pathogen removal by 'NATURAL BIOCONTROL and DISINFECTION' of treated wastewater by predator microbes. Pathogens may also be adsorbed to finer particles and sediment. Pathogen removal occurs in several ways-

1   Natural UV radiation

2   Chemical oxidation
3   Attack by lytic bacteria and bacteriophages (viruses)
4   Filtration, sedimentation, absorption, and
5   Natural die-off.

Both sub-surface flow wetland (SSF) and surface flow wetland (FWS) can achieve up to 90 % removal of fecal coliform bacteria from primary treated sewage and 99 % removal from secondary treated sewage. Study by QDNR (2000) have shown that constructed wetlands can remove 95 % of pathogens and indicator organisms. Fecal-coliform removal is also very high, producing effluent with < 1000 cfu/100 mL and as low as 100 cfu/100 mL, acceptable for crop and golf course irrigation and agriculture use.

In order to maximize pathogen removal, a combination of densely vegetated and open water zones should be used. The densely vegetated zones maximize filtration and sedimentation of particles to which pathogens will be adsorbed, while the open water zones will maximize UV disinfection. Healthy populations of natural-wetland microbes (bacteria and viruses) should be encouraged to promote predation, lysis and competition with pathogenic human microbes. Water fleas like *Daphnia* can predate upon bacteria and phytoplankton.

APPLICATIONS OF CONSTRUCTED WETLAND TECHNOLOGY TO TREAT INDUSTRIAL WASTEWATER

Applications of constructed wetland technology to treat industrial wastewater are increasing. They include-

1   1). Treatment of food processing wastewater e.g. fruit and vegetables, sugar production, poultry and meat processing, breweries and distilleries;
2   2). Treatment of textile wastewater & paper and pulp mill wastewater;
3   3). Treatment of wastewater from petrochemicals e.g. polishing of secondarily treated refinery wastewater and wash-down runoff from petrochemical industries;
4   4). Treatment of mining wastewater e.g. acid coal mine drainage with high concentrations of dissolved iron, manganese, aluminum and sulfate; metal mine drainage from lead, zinc, silver, copper, nickel and uranium mines & cyanide from gold & silver mining.

Some Success Stories of Treatment of Industrial Wastewater by Aquatic Plants in Constructed Wetlands

Reports about full scale applications of constructed wetlands in removal of xenobiotics or recalcitrant compounds from industrial effluents are scarce. There are few reports-

Treatment of Chemical Industry Effluents

Dias (2000), reported about 'Vertical Flow Reed (*Phragamites* Spp) Bed' installed at the chemical industry Quimigal, S.A. in Portugal. This constructed wetland has total planted area of 10,000 sqm, which efficiently removes toxic nitro-aromatic compounds, such 'aniline', 'nitro-phenols' and 'nitrobenzene'.

Treatment of Textile Wastewater

Industrial textile dyes are important component in the textile wastewater. Textile dyes have been designed and synthesized to be highly persistent and resistant to washing and action of chemical solvents and sunlight. There are currently more than 10,000 different textile dyes commercially available in the world markets. Azo, indigoid and anthraquinone are the major chromophores used in the textile industries. They are complexed with heavy metals copper (Cu), cobalt (Co) and chromium (Cr) and are of considerable public health concern. Half-life of some of the textile dyes like the 'reactive blue' 19 is 46 years at 25^0 C and pH 7. Reactive dyes typically have poor fixation rates to fabrics, and dye concentrations up to 1,500 mg/L could be found in the liquor that is discharged into the sewers.

One of the first experiment made was in the 'Horizontal Bed Reed (*Phragamites* Spp.) of 150 sqm in Australia (Davis & Cottingham, 1994). In Georgia, U.S., Coats American is currently using constructed wetlands as the final step in the textile wastewater treatment operations. There is remission of a large portion of residual dye content, an appreciable decrease in the chemical oxygen demand (COD), and apparent lower chronic toxicity from the textile effluent. (Dias, et al., 2006). In addition to color removal the wetlands plants also removed the heavy metals from the dyestuff by bio-accumulation

Treatment of Cyanide Contaminated Gold Mine
Wastewater in Constructed Wetlands

Cyanide is the leach reagent of choice for gold (Au) and silver (Ag) extraction during gold & silver mining and consequently results in the mined wastewaters. A diluted sodium cyanide (0.05 %) solution is sprayed on gold-containing crushed ore, placed in heaps. The cyanide readily forms a water-soluble complex with the gold from which the precious metal is recovered. Currently there are about 875 gold and silver mines throughout the world, of which about 460 use cyanide as leaching reagent, thus using 347,000 tons of sodium cyanide every year contaminating huge volume of wastewater. (Prasad, 2006).

Salix spp. & *Sorghum* spp have been successfully used in constructed wetlands to detoxify the cyanide contaminated wastewaters. These plants possess the cyanide detoxifying enzyme system called 'beta-cyanoalanine synthase' which phytotransform cyanide into 'aspargine', a non-toxic essential amino acids in plants. (Prasad, 2006).

COST OF COMBATING GROUNDWATER POLLUTION BY PLANTS

Current groundwater cleanup technologies, such as 'granular activated carbon' and 'advanced oxidation methods' incur heavy expenditure and are cost-prohibitive. After the Chernobyl disaster in Ukraine in 1986, the estimated cost of removal of radionuclides from contaminated pond waters was calculated to be US $ 80 per thousand gallons by physical process. The phytoremediation technology using sunflower ranged from about US $ 2 – $ 6 per thousand gallons of water.

The US Army Environmental Centre is developing cost-effective phytoremediation technologies by constructed wetlands to cleanup groundwater contaminated with residues of explosives like TNT, RDX, DNT and Octahydro-1,3,5,7- tetranitro-1,3,5,7-tetraazocine (HMX).

CONCLUSIONS

Plant based technologies (phytoremediation technologies) to combat environmental pollutants of air and water are very cost-effective, economically viable, ecologically compatible and socially acceptable methods for management of polluted air and waters as compared to the physical and chemical methods.

Air pollutants like sulfur and nitrogen oxides, ozone and suspended particulate matters (SPMs) can also be ameliorated by plants. Despite of the adverse effects of pollutants on plant life there are some tolerant 'wonder species' which can absorb, adsorb, detoxify, bio-accumulate & metabolise the pollutants. They act as a 'living filter and natural sink' for the air pollutants. Researches into biodiversity and environmental biotechnology has helped in identifying and producing more natural pollution fighters.

Over the past 20 years the constructed wetland technology using aquatic plant species specially the reeds (*Phragmites* Spp.) and the cattails (*Typha* Spp.) has been widely accepted in Europe and America as a low-cost, environmentally sustainable option to the conventional wastewater treatment facilities for both municipal and industrial wastewaters.

Phytoremediation of polluted air & water can be implemented by several naturally occurring plants whose numbers are growing after studies identify and more and more such plants.

REFERENCES & ADDITIONAL READINGS

I. PLANTS & PHYTOREMEDIATION TECHNOLOGY COMBATING AIR POLLUTION

Anonymous (1986): Pollution on Plants; *Science Reporter*, Pub. Of Council of Scientific & Industrial Research (CSIR), New Delhi; March, 1986; pp. 189-193.

Agarwal, M. Singh, SK,. Singh, J., & Rao, DN (1991): Biomonitoring of Air Pollution Around Industrial Sites; *J. of Environmental Biology;* pp. 211-222.

Agarwal, S.B. and Madhulika Agarwal (2000); *Environmental Pollution and Plant Responses* (Ed.); CRC / Lewis Publisher, USA.

Bergmann, E., Bender, & J., Weigel HJ (1995): Growth Response and Foliar Sensitivities of Native Herbaceous Species to Ozone Exposure; Water, *J. of Air and Soil Pollution,* Vol. 85: pp. 1437-1442.

Dochinger, L.S. (1980): Interception of Air Borne Particulates by Tree Planting; *J. of Environmental Quality;* Vol. 9: pp. 265-268.

Freer, PH-S & El AA-K, Taylor, G. (2004): Capture of Particulate Pollution by Trees : A Comparison of European & N. American Species; *J. of Air and Soil Pollution,* Vol. 155: pp. 173 – 187.

Hill, A.C. (1971): Vegetation : A Sink for Atmospheric Pollutants; *J. of Air Pollution and Control Association,* Vol. 21: pp. 341-346.

Izuta, T., Miyake, H., & Totsuka, T. (1993): Evaluation of Air-Polluted Environment Based on Growth of Radish Plants Cultivated in Small Sized Open-top Chambers; *J. of Environmental Science,* Vol. 2: pp. 25-37.

Khan, A.M., Pandey V., Yunus, M., & Ahmad, K.J. (1989): Plants as Dust Scavengers : A Case Study; *The Indian Foresters*; Vol. 115 (9): pp. 670-672.

Manning, WJ., & Feder, WA. (1980): Biomonitoring of Air Pollutants With Plants; Applied Science Publishers, London.

Mansfield, TA., Free-Smith, PH (1984): The Role of Stomata in Resistance Mechanisms; In Kozoil M.J. & Whatley, F.R. (eds.) (1984): *Gaseous Air Pollutants and Plant Metabolism;* pp. 131-146; London Butter Worths.

Nali, C., Crocicchi, L., & Lorenzini, G. (2004): Plants as Indicator of Urban Air Pollution (Ozone & Trace Elements) in Pisa, Italy; *J. of Environmental Monitor,* Vol. 6: pp. 636-645.

Nasurallah, M., Tatsumoto, H., & Misawa, A. (1994): Effect of Roadside Planting on Suspended Particulate Matters (SPM) Concentration Near Road; *J. of Environmental Technology,* Vol. 15: pp. 293-298.

Okano, K.& Totsuka, T. (1986): Absorption of Nitrogen Dioxide by Sunflower Plants Grown at Various Levels of Nitrate; *New Phytology,* Vol. 102; pp. 551-556.

Orwell, RL., Ronald, L., Wood, RL., Taran, J., Torpy, F., & Burchet, MD. (2004): Removal of Benzene by the Indoor Plant & Implications for Air Quality; *J. of Water, Soil & Air Pollution,* Vol. 157: pp.193-207

Posthumus, A.C. (1983): Higher Pants as Indicators and Accumulators of Gaseous Air Pollutants; *J. of Environmental Monitor & Assessment;* Vol. 3; pp.263-272.

Simonich, S.L. & Hites, R.A. (1994): Importance of Vegetation in Removing Polycyclic Aromatic Hydrocarbons (PAHs) from the Atmosphere; *Nature,* Vol. 370; pp. 49-51.

Singh, N., Yunus, M., Srivastva, K., Singh, S.N., Pandey, V., Misra, J., & Ahmad, K.J. (1995): Monitoring of Auto Exhaust Pollution by Roadside Plants; *J. of Environmental Monitor & Assessment;* (USA), Vol. 34; pp. 13-25.

Singh, S.N. and Amitosh Verma (2006): *Phytoremediation of Air Pollutants : A Review*; In S.N. Singh & R.D. Tripathi (ed.) '*Environmental Bioremediation Technologies*'; Springer Publication, New York; pp. 275-292.

Singh, S.K. & Rao, D.N. (1983): Evaluation of Plants for Their Tolerance to Air pollution; In Mathur, H.B. & Pal, K. (eds.) 'Proceedings of Symposium on Air Pollution Control; IIT Delhi; pp. 218-224.

Takahashi, M., Sasaki, Y., Ida, S., & Morikawa, H. (2001): Nitrate reductase gene enrichment improves assimilation of nitrogen dioxide in Arabidopsis; *J. of Plant Physiology*, Vol. 126; pp. 731-741.

Upadhaya, J.K. & N. Kobayashi (2006): *Phytomonitoring of Air Pollutants for Environmental Quality Management*; In S.N. Singh & R.D. Tripathi (ed.) '*Environmental Bioremediation Technologies*'; Springer Publication, New York; pp. 293 - 314.

Varshney, C.K. & Mitra, I. (1993): Importance of Hedges in Improving Urban Air Quality; *J. of Landscape & Urban Planning*, Vol.25; pp. 75-83.

Verma, A. (2003): Attenuation of Automobile Generated Air Pollution by Higher Plants; Unpublished Ph.D Thesis, University of Lucknow, India.

Wolverton, B.C, McDonald, R.C., & Mesick, H.H. (1985): Foliage Plants for the Indoor Removal of the Primary Combustion Gases Carbon Monoxide and Nitrogen Oxides; *J. of Missi Academy of Sciences;* Vol. 30; pp. 1-8.

Wolverton, B.C & Wolverton, J.D. (1993): Plants & Soil Microorganisms – Removal of Formaldehyde, Xylene and Ammonia from the Indoor Environment; *J. of Missi Academy of Sciences;* Vol. 38, pp. 11-15.

Wolverton, B.C. (1997): *How to Grow Fresh Air : Fifty Houseplants That Purify Your Home or Office*; Penguin Books, NY; First Published in UK as '*Eco-friendly Houseplants*', Weidenfeld & Nicolson, London (1996).

Yunus, M. & Iqbal, M. (eds.) (1996): *Plant Response to Air Pollution*; John Wiley, UK., p. 1-34

II. PLANTS & PHYTOREMEDIATION TECHNOLOGY COMBATING WATER POLLUTION

Alvarado, S., Guedez, M, Marco P,. L. M., Graterol, N., Anzalone, A., Arroyo, J. and Zaray, Gy (2008): 'Arsenic removal from waters by bioremediation with the aquatic plants water hyacinth *(Eichhornia crassipes)* and Lesser Duckweed *(Lemna minor)*; *J. of Bioresource Technology*; Vol. 99: pp. 8436 - 8440.

Baker, K.H. & Herson, D.S. (1994): *Bioremediation*; McGraw Hill, Inc.,New York

Barber, J.T.; Sharma, H.A.; Ensley, H.E.; Polito, M.A.; and Thomas D.A. (1995): Detoxification of Phenol by the Aquatic Angiosperm *Lemna gibba; Chemosphere*; Vol. 31: 6; p. 3567 –3568.

Baetens, T (1993) : Wastewater Recycling by Using Aquatic Plants; In *Proceedings of the Awareness Workshop for a Sustainable Future*; Centre for Scientific Research, Auroshpillam, India; Nov. 28-Dec.5, 1993.

Betts, K.S. (1997): Native Aquatic Plants Remove Explosives; *Journal of Environmental Science & Technology*, Vol. 31: pp. 304 A

Brooks, R.R. (ed.) (1998): *Plants That Hyperaccumulates Heavy Metals*; Cambridge University Press.

Bulusu, K.R. and Pande, S.P., (1990): Nitrates – A Serious Threat to Groundwater Pollution; *BHU JAL News (Quarterly Journal of Central Groundwater Board)*; Vol. 5 (2): pp. 39.

Cave, Shane (1991): A Green Revolution Down at the Sewer Ponds (Sewage Purification by Duckweeds (*Lemna* Spp.); *Our Planet*; UNEP Publication, Nairobi, Kenya; Vol.3(1); pp. 12-13;

Cunningham, S.D. and Ow D.W. (1996): Promises and Prospects of Phytoremediation; *Plant Physiology*; Vol. 110; pp. 715-719.

Campbell, W.H. (1996): Nitrate Reductase Biochemistry Comes of Age; *J. of Plant Physiology*, Vol. 111: pp. 355-364.

Davies, T.H. & Cottingham, P.D. (1994): The Use of Constructed Wetlands for Treating Industrial Effluents (Textile Dyes); *J. of Water Science & Technology*, Vol. 29: pp. 227-232.

Dinges, R (1978): Upgrading Stabilization Pond Effluents by Water Hyacinth (*Eichhornia cressipes*) Culture; *Journal of Water Pollution Control Federation*; US; Vol. 50; pp. 833-845.

Dias, S.M. (2000): *Nitroaromatic Compound Removal in a Vertical Flow Reed Bed : Case Study Industrial Wastewater Treatment*; Intercost Workshop on Bioremediation, Sorrento, Italy; pp. 119-120.

Dushenkov, S., Vasudev, D., Kapulnik, Y., Gleba, D., Fleisher, D., Ting, KC., & Ensley, B. (1997): Removal of Uranium from Water Using Terrestrial Plants' *J. of Environmental Science & Technology*, Vol. 31; pp. 3468-3474.

Dwivedi, U.N., Seema Mishra, Poorinima Singh & R.D. Tripathi (2006): *Nitrate pollution and its Remediation*; In S.N. Singh & R.D. Tripathi (ed.) '*Environmental Bioremediation Technologies*'; Springer Publication, NY.; pp. 353-389.

Eapen, Susan., Singh, Shraddha., & D'Souza, SF (2006): *Phytoremediation of Metals & Radionuclides*; In: S.N. Singh & R.D. Tripathi (eds.) '*Environmental Bioremediation Technologies*'; Springer Publication, NY.; pp. 189-209.

Evans, J (1991): Safe Drinking Water for the Developing World; *Our Planet*; Vol. 3, No. 1; pp. 12-13; UNEP Pub., Nairobi, Kenya.

Eweis, J.B., Ergas, S.J., Chang, D.P. and Schroeder, E.D. (1998): *Bioremediation Principles*; McGraw Hill, Boston, USA.

Flathman, P.E., Jerger, D.E. and Exner, J.H. (eds).(1994): *Bioremediation : Field Experience;* Lewis Publishers, Boca Raton.

Gopal, B. (1987): *Water Hyacinth*; Elsevier Science Publishers; Amsterdam.

Greenway, Margaret (2000): *Constructed Wetlands as Ecologically Sustainable Options for Water Pollution Control: A Challenge for Environmental Engineers and Environmental Technologists;* Griffith University, Brisbane, Australia.

Jayaraman, AP & Prabhakar, S. (1982): *The Water Hyacinth Uptake of Cesium (Cs) and Strontium (Sr) and its Decontamination Potential as an Approach to the Zero Release Concept;* In: *Proceedings Of International Symposium* on 'Migration in the Terrestrial Environment of Long-lived Radionuclides from the Nuclear Fuel Cycle'; Knoxville T.N. International Atomic Energy Agency, Vienna, Austria.

Joglekar, VR and Sonar, VG (1987): Application of Water Hyacinth for Treatment of Domestic Water, Generation of Biogas and Organic Manure; In KR Reddy and WH Smith (eds.) *'Aquatic Plants for Water Treatment and Resource Recovery'*; Magnolia Pub. Inc.; Orlando, USA; pp. 747-753.

Kamely, Daphne; Ananda Chakarbarthy and Gilbert S. Omenn (1990): *Biotechnology and Biodegradation*; Gulf Publishing Company from the New York Environmental Facilities Corp., NY 12205.

Kamal, M., Ghaly, A.E., Mahmoud, N. & Cote, R. (2004): Phytoaccumulation of Heavy Metals by Aquatic Plants; *J. of Environmental International*; Vol. 29: pp. 1029 – 1039.

Kumar, P and Garde, RJ (1994): *Upgrading Wastewater Treatment by Water Hyacinth in Developing Countries*; University of Roorkee Pub.; Roorkee, India.

Low, K.S., Lee C.K., and Tan K.K. (1995): Biosorption of Basic-Dyes by Water Hyacinth Roots; *J. of Bioresource Technology*, Vol. 52: pp. 79-83.

Manning, K. (1988): Detoxification of Cyanide by Plants and Hormone Action; In *'Cyanide Compounds in Biology'* (ed.); CIBA Foundation, John Wiley & Sons, UK.

NAS (1977): *Making Aquatic Weeds Useful : Some Perspective for Developing Countries*; National Academy of Sciences Pub.; Washington DC, USA.

Nicholas, P. Cherimisinoff (1996): *Biotechnology for Waste and Wastewater Treatment*; Noyes Publication

Prasad, M.N.V. (2006): Aquatic Plants for Phytotechnology; In S.N. Singh & R.D. Tripathi (ed.) *'Environmental Bioremediation Technologies'*; Springer Publication, NY.; pp. 259 – 274.

Reddy, KR and Sutton, DL (1984): Water hyacinth *(Eichhornia cressipes)* for water quality improvement and biomass production; *Journal of Environmental Quality*; USA; Vol.13(1); pp. 1-8.

Sinha, Rajiv K (1996): Sewage as a resource : A case study of afforestation using sewage irrigation in Jaipur, Rajasthan, India: *The Environmentalist*, U.K; Vol. 16: pp. 91 - 94.

Sinha, Rajiv K (1997): River pollution in India : Innovative technologies for sewage treatment under the Ganga and Yamuna Action Plans; *International Journal of Environmental Education and Information;* University of Salford, UK: Vol. 16(4): pp.395-406.

Sinha, Ambuj K and Rajiv K. Sinha, (2000): Sewage Management by Aquatic Weeds (Water Hyacinth and Duckweed): Economically Viable and Ecologically Sustainable Bio-mechanical Technology; *International Journal of Environmental Education and Information*; University of Salford, UK; Vol. 19: No. 3;

Sinha, Rajiv K., Sunil Herat and P.K. Tandon (2003): Phytoremediation : A Cost-effective, Ecologically Sustainable and Socially Acceptable Bioengineering Technology for Rehabilitation of Contaminated and Eroded Lands (Sites) and Purification of Polluted Water by the Use of Vetiver Grass *(Vetiveria zizanioides* Linn).; *Proceedings of the 'National Environment Conference';* June 18-20, 2003, Brisbane Convention Center, Brisbane, QLD, Australia .

Schnoor, J.L., Licht, L.A., McCutcheon, S.C., Wolfe, N.L. & Carreira, L.H. (1995): Phytoremediation of Organic and Nutrient Contaminant; *J. Of Environmental Science & Technology*; Vol. 29: pp. 318-323.

Sriprang, R,. & Murooka, Y. (2006): Accumulation & Detoxification of Metals by Plants & Microbes; In S.N. Singh & R.D. Tripathi (ed.) *'Environmental Bioremediation Technologies'*; Springer Publication, NY.; pp.77-100.

Stowell, R; Ludwig, R; Colt, J and Tchobanoglous, G (1981): Concepts in aquatic plant systems treatment design; *Journal of Environmental Engineering Division*; American Society of Civil Engineers; Vol. 107 (5); pp.919-940.

Suthersan, S.S. (1997): *Remediation Engineering : Design Concepts;* Geraghty & Miller, CRC Inc. Florida, USA.

Steve, Prentis (1984): *Biotechnology- A New Industrial Revolution*; Orbis Publishing House.

Susarla, S., Bachhus, T.S., Wolfe, N.L., & McCutcheon, C.S. (1999): Phytotransformation of Perchlorate Using Parrot Feather; *Soil and Groundwater Cleanup*, Vol. 2: pp. 20-23.

Susarla, S., Medinia, V.F. and McCutcheon S.C. (2002): Phytoremediation : An ecological Solution to Organic Chemical Contamination; J. of Ecological Engineering, Vol. 18: pp. 647 – 658.

Tchobanoglous, G (1987): Aquatic plant systems for waste water treatment engineering considerations; In KR Reddy and WH Smith (eds.) *'Aquatic Plants for Water Treatment and Resource Recovery'*; Magnolia Pub. Inc.; Orlando, USA; pp. 27-48.

Tripathi, BD and Shukla SC (1991): Biological Treatment of Wastewater by Selected Aquatic Plants; *Journal of Environmental Pollution*; Vol. 69 (Pub. In WWF (India), New Delhi, 1991).

Terry, M. (2003): Phytoremediation of Heavy Metals from Soils; *Advances in Biochemical Engineering / Biotechnology*; Vol. 78: pp. 97-123.

UNEP (1991) : A Green Revolution Down at the Sewer Ponds; *Our Planet*, Vol. 3. No. 1; pp. 12-13; UNEP Pub., Nairobi, Kenya (Shane Cave).

USEPA (1996): *A Citizen's Guide to Phytoremediation*; United States Environmental Protection Agency; htpp://clu-in-com/citguige/phyto.htm

Waranusantigul, P., Pokethitiyook, P., Kruatrachue, M., and Upatham, E.S. (2003): Kinetics of Basic Dye (Methylene Blue) Biosorption by Giant Duckweed (*Spirodela polyrrhiza*); *J. of Environmental Pollution*, Vol. 125: pp. 385-392.

Weis, J.S. & Weis, P. (2004): Metal Uptake, Transport and Release by Wetland Plants: Implications for Phytoremediation and Restoration; *J. of Environmental International*, Vol. 30: pp. 685 – 700.

WWF (1991): *The Wetlands of Calcutta : Sustainable Development or Real Estate Takeover (Report about use of water hyacinth in wastewater purification)*; In WWF (India) Pub. New Delhi.

Zheng, Chun Rong; Tu, Cong; and Chen Huai Man (1998): Preliminary Experiment on Purification of Eutrophic Water with Vetiver; *Proceedings of International Vetiver Workshop*, Fuzhou, China, Oct. 21-26, 1997.

III. Plants & Phytoremediation Technology Combating Soil Pollution

Aboulroos, S.A., Helal, M.I.D., & Kamel M.M. (2006) : Remediation of Pb and Cd Polluted Soils Using In-Situ Immobilization and Phytoextraction Techniques ; *J. of Soil and Sediment Contamination* ; Vol. 15 : pp. 199- 215.

Baker, A.J.M., & Walker, P.L. (1989) : Ecophysiology of Metal Uptake By Tolerant Plants ; In A. Shaw (Ed.) *Heavy Metal Tolerance in Plants – Evolutionary Aspects*; pp. 155 – 177 ; CRC Press

BIO-WISE (2000) : *Contaminated Land Remediation : A Review of Biological Technology* ; London, DTI.

Fulekar, M.H. & Jadia, C.D. (2008) : Phytoremediation : The Application of Vermicompost to Remove Zn, Cd, Cu, Ni, and Pb by Sunflower Plants ; *Environmental Engineering and Management Journal* ; Vol. 7 (5) : pp. 547 – 558.

Kuzovkina, Y.A., Knee, M., & Quigley, M.F. (2004) : Cadmium and Copper Uptake and Translocation in Five Willow (*Salix*) Species ; *International J. of Phytoremediation* ; Vol. 6 : pp. 269 – 287.

Lasat, M.M. (2001) : Phytoextraction of Toxic Metals ; A Review of Biological Mechanisms ; *Journal of Environmental Quality* ; Vol. 31 : pp. 109 – 120.

Peng, K., Li, X., Luo, C., & Shen, Z. (2006) : Vegetation Compostition and Heavy Metal Uptake by Wild Plants at Three Contaminated Sites in Xiangxi Area, China ; *Journal of Environmental Science and Health* (Part A) ; Vol. 40 : pp. 65 – 76.

Qui, R., Fang, X., Tang, Y., Du, S., & Zeng, X. (2006) : Zinc Hyperaccumulation and Uptake by *Potentilla Griffithi* Hook. ; *International J. of Phytoremediation* ; Vol. 8 : pp. 299 – 310.

Suzuki, N. & Koizumi, N. (2001) : Screening of Cadmium (Cd) Responsive Genes in *Arabidopsis thaliana ; Plant Cell Environment* ; Vol. 24 : pp. 1177 – 1188.

Zorpas, A.A., Arapoqlou, D., Panaqiotis, K. (2003) : *Soil Phytoremediation* ; Paper Presented at the 4th International Conference on Engineering Geotechnology ; Rio de Janeiro, Brazil.

USEFUL WEBSITES

1). Nitrate in News : http:// www.nitrate.com.nitrate1.html
2). United States Environmental Protection Agency; htpp://clu-in-com/citguige/phyto.htm
3). Shewanella Federation : www.shewanella.org/why_shewanella.sjsp

INDEX